NISTIR 85-3273-28

Energy Price Indices and Discount Factors for Life-Cycle Cost Analysis – 2013
Annual Supplement to NIST Handbook 135 and NBS Special Publication 709

Amy S. Rushing
Joshua D. Kneifel
Barbara C. Lippiatt
Applied Economics Office
Engineering Laboratory

June 2013

U.S. Department of Commerce
Cameron F. Kerry, Acting Secretary of Commerce

National Institute of Standards and Technology
Patrick D. Gallagher, Under Secretary of Commerce for Standards and Technology and Director

ABSTRACT

This is the 2013 edition of energy price indices and discount factors for performing life-cycle cost analyses of energy and water conservation and renewable energy projects in federal facilities. It will be effective from April 1, 2013 to March 31, 2014. This publication supports the federal life-cycle costing methodology described in 10 CFR 436A and OMB Circular A-94 by updating the energy price projections and discount factors that are described, explained, and illustrated in NIST Handbook 135 (HB 135, Life-Cycle Costing Manual for the Federal Energy Management Program.) It supports private-sector life-cycle cost analysis by updating the energy price indices that are described, explained, and illustrated in NBS Special Publication 709 (SP 709).

Disclaimer:
Certain trade names or company products are mentioned in the text to specify adequately the software and operating systems used for performing the life-cycle cost analyses. In no case does such identification imply recommendation or endorsement by the National Institute of Standards and Technology, nor does it imply that the software and operating systems are the best available for the purpose.

PREFACE

This is the 2013 Annual Supplement to NIST Handbook 135, Life-Cycle Costing Manual for the Federal Energy Management Program (FEMP). The annual supplement provides energy price indices and discount factors for use with the Federal Energy Management Program's procedures for life-cycle cost analysis, as established by the U.S. Department of Energy (DOE) in Subpart A of Part 436 of Title 10 of the Code of Federal Regulations (10 CFR 436A), and amplified in NIST Handbook 135. These indices and factors are provided as an aid to implementing life-cycle cost evaluations of potential energy and water conservation and renewable energy investments in existing and new federally owned and leased buildings.

The price indices and discount factors are calculated with the most recent energy price projections from DOE's Energy Information Administration (EIA) and the most recent discount rates from FEMP and the Office of Management and Budget (OMB) Circular A-94. This issue of the Annual Supplement is intended for use from April 1, 2013 to March 31, 2014. The updated edition of the NIST Building Life-Cycle Cost (BLCC) and Energy Escalation Rate Calculator (EERC) programs are released at the same time as this Annual Supplement, for use over the same time period. The software products are discussed below.

At the request of a number of users, a text file of the EIA energy price projections underlying this Annual Supplement has been made available by NIST. To obtain this file (ENCOST13.txt), please visit the DOE/FEMP web site at http://www1.eere.energy.gov/femp/program/lifecycle.html.

The life-cycle costing methods and procedures, as set forth in 10 CFR 436A, are to be followed by all federal agencies, unless specifically exempted, for evaluating the cost effectiveness of potential energy and water conservation and renewable energy investments in federally owned and leased buildings. For most other federal LCC analyses OMB Circular A-94 provides the relevant guidelines.

As called for by legislation (Energy Policy and Conservation Act, P.L.94-163, 1975, 92 Stat 3206, 42 USC 8252 et seq), the National Institute of Standards and Technology has provided technical assistance to the U.S. Department of Energy in the development and implementation of life-cycle costing methods and procedures. The following publications and software products provide the methods, data, and computational tools for federal life-cycle cost analysis:

(1)   *Life-Cycle Costing Manual for the Federal Energy Management Program*, National Institute of Standards and Technology, Handbook 135 (1995).

This manual is a guide to understanding life-cycle costing and related methods of economic analysis as they are applied to federal decisions, especially those subject to 10 CFR 436A rules. It describes the required procedures and assumptions, defines and explains how to apply and interpret economic performance measures, gives examples of federal decision problems and their solutions, explains how to use energy price indices and discount factors, and provides worksheets and other computational aids and instructions for calculating the required measures.

(2)   *Energy Price Indices and Discount Factors for Life-Cycle Cost Analysis, Annual Supplement to NIST Handbook 135 and NBS Special Publication 709*, National Institute of Standards and Technology, NISTIR 85-3273.

This report, which is updated annually, provides the current DOE and OMB discount rates, projected energy price indices, and corresponding discount factors needed to estimate the present values of future energy and non-energy-related project costs.

(3)   *BLCC 5.3-13, NIST Building Life-Cycle Cost.* This program uses as default values the same discount factors and energy price projections that underlie the discount factor tables in the Annual Supplement. It is available for Windows, Mac OS X, and Linux.

BLCC 5.3 provides comprehensive economic analysis capabilities for the evaluation of proposed capital investments that are expected to reduce the long-term operating costs of buildings and building systems. It computes the LCC for project alternatives, compares project alternatives in order to determine which has the lowest LCC, performs annual cash flow analysis, and computes net savings (NS), savings-to-investment ratio (SIR), and adjusted internal rate of return (AIRR) for project alternatives over their designated study period. The BLCC program can be used to perform economic analysis of capital investment projects undertaken by federal, state, and local government agencies. In the application to federal energy conservation and renewable energy projects, BLCC5 is consistent with NIST Handbook 135, and the federal life-cycle cost methodology and procedures described in 10 CFR 436A and OMB Circular A-94.

The BLCC5 User's Guide is part of its Help system. BLCC 5.3 has six modules, all of them consistent with the life-cycle cost methodology of 10 CFR 436A, but programmed to include default inputs and nomenclature for specific uses:

- **FEMP Analysis, Energy Project**
  for energy and water conservation and renewable energy projects under the FEMP rules, agency-funded;

- **Federal Analysis, Financed Project**
  for federal projects financed through Energy Savings Performance Contracts (ESPC) or Utility Energy Services Contracts (UESC) as authorized by Executive Order 13123 (6/99);

- **OMB Analysis, Federal Analysis, Projects subject to OMB Circular A-94**
  for projects subject to OMB Circular A-94 (most other, non-energy, federal government construction projects, but not water resource projects);

- **MILCON Analysis, Energy Project**
  for energy and water conservation and renewable energy projects in military construction, agency-funded;

- **MILCON Analysis, ECIP Project**
  for energy and water conservation projects under the Energy Conservation Investment Program (ECIP).

- **MILCON Analysis, Non-Energy Project**
  for military construction designs that are not primarily for energy or water conservation.

(4)   *EERC 2.0-13, Energy Escalation Rate Calculator*, a program that computes an average rate of escalation for a specified time period, which can be used as an escalation rate for contract payments in Energy Savings Performance Contracts (ESPC) and Utility Energy Services Contracts (UESC). Escalation rates can be computed based on the EIA energy price projections used for calculating the FEMP discount factors and on EIA projections adjusted by NIST for potential carbon pricing.

The latest versions of the programs and publications described above can be downloaded from the DOE/FEMP web site at http://www1.eere.energy.gov/femp/program/lifecycle.html.

The U.S. Department of Energy was directed by legislation and executive order to make available to the private sector the methods, procedures, and related aids developed for federal use. In response to this directive, the National Institute of Standards and Technology, under sponsorship of the U.S. Department of Energy, published a life-cycle costing book for use by the private sector entitled Comprehensive Guide for Least-Cost Energy Decisions, NBS SP 709 (January 1987).

In 2008, DOE and NIST presented a 2½-hour webcast, "Overview of Federal Building Energy Efficiency Mandates/An Introduction to Building Life-Cycle Costing." This free webcast that introduces the elements of life-cycle cost analysis of energy and water conservation projects is available at http://www.energycodes.gov/training-courses/overview-federal-building-energy-efficiency-mandates. For in-house training, FEMP-Qualified Instructors are available to conduct LCC workshops on their own account across the U.S. For a list of instructors e-mail joshua.kneifel@nist.gov.

For further information on the Federal Energy Management Program, please visit http://www1.eere.energy.gov/femp.

ACKNOWLEDGMENTS

The authors wish to thank Cyrus Nasseri of the Federal Energy Management Program, U.S. Department of Energy (DOE), for his support and direction of this work. Appreciation is extended to Paul Kondis and Paul Holtberg, of the DOE Energy Information Administration, for providing the energy price projections upon which this report is based. Thanks are also due to Stephen Petersen and Sieglinde Fuller, who originated this publication.

CONTENTS

LIST OF TABLES

ABBREVIATIONS

A	-	Annual amount
A_0	-	Annual amount at base-date prices
ADAGE	-	Applied Dynamic Analysis of Global Economy
AEO2013	-	Annual Energy Outlook 2013 (DOE-EIA publication)
BLCC	-	NIST Building Life Cycle Cost computer program
CO_2	-	Carbon Dioxide
COAL	-	Coal
d	-	discount rate
DIST	-	Distillate Oil
DOE	-	U.S. Department of Energy
e	-	price escalation rate (annual rate of price change)
EIA	-	Energy Information Administration (DOE)
ELEC	-	Electricity
EPA	-	U.S. Environmental Protection Agency
ESPC	-	Energy Savings Performance Contract
FEMP	-	Federal Energy Management Program
FY	-	Fiscal Year
GASLN	-	Gasoline
kg	-	kilogram
LCC	-	Life-Cycle Cost
LPG	-	Liquefied petroleum gas
N	-	Number of discount periods (in years)
NEMS	-	National Energy Modeling System
NIST	-	National Institute of Standards and Technology
NTGAS	-	Natural Gas
OMB	-	Office of Management and Budget
RESID	-	Residual Oil
SPV	-	Single Present Value (factor)
UESC	-	Utility Energy Services Contract
UPV	-	Uniform Present Value (factor)
UPV^*	-	Modified Uniform Present Value (factor)

INTRODUCTION

This report provides tables of present-value factors for use in the life-cycle cost analysis of capital investment projects for federal facilities. It also provides energy price indices based on Department of Energy (DOE) forecasts from 2013 to 2043. The factors and indices presented in this report are useful for determining the present value of future project-related costs, especially those related to operational energy costs. Discount factors included in this report are based on two different federal sources: (1) the DOE discount rate for projects related to energy conservation, renewable energy resources, and water conservation; and (2) Office of Management and Budget (OMB) discount rates from Circular A-94 for use with most other capital investment projects in federal facilities.

The DOE discount and inflation rates for 2013 are as follows:

Real rate (excluding general price inflation):	3.0 %
Nominal rate (including general price inflation):	2.5 %
Implied long-term average rate of inflation:	-0.5 %

The DOE nominal discount rate is based on long-term Treasury bond rates averaged over the 12 months prior to the preparation of this report. The nominal, or market, rate is converted to a real rate to correspond with the constant-dollar analysis approach used in most federal life-cycle cost (LCC) analyses. The method for calculating the real discount rate from the nominal discount rate is described in 10 CFR 436 and uses the projected rates of general inflation published in the most recent Report of the President's Economic Advisors, Analytical Perspectives. The procedure would result in a discount rate for 2013 lower than the 3.0 % floor prescribed in 10 CFR 436. Thus the 3.0 % floor is used as the real discount rate for FEMP analyses in 2013. The implied long-term average rate of inflation was calculated as -0.5 %. Federal agencies and contractors to federal agencies are required by 10 CFR 436 to use the DOE discount rates when conducting LCC analyses related to energy conservation, renewable energy resources, and water conservation projects for federal facilities.

The nominal and real discount rates applicable to general (non-energy or water) capital investments are published annually in OMB Circular A-94, Appendix C. OMB has specified two basic types of discount rates: (1) a discount rate for public investment and regulatory analyses; and (2) a discount rate for cost-effectiveness, lease-purchase, and related analyses. Only discount rates for the second type of analyses are included in this Annual Supplement, since the primary purpose of this report is to support cost-effectiveness studies related to the design and operation of federal facilities.

OMB discount rates for cost-effectiveness and lease-purchase studies are based on interest rates on Treasury Notes and Bonds with maturities ranging from 3 to 30 years. Currently (as of April 2013) six maturities have been specifically identified by OMB, and are shown here with the corresponding real interest rate to be used as the discount rate for studies subject to OMB Circular A-94:

Maturity:	3-year	5-year	7-year	10-year	20-year	30-year
Rate:	-1.4 %	-0.8 %	-0.4 %	0.1 %	0.8 %	1.1 %

OMB suggests that the actual discount rate for an economic analysis be interpolated from these maturities and rates, based on the study period used in the analysis. Due to limitations on the size of this Annual Supplement, discount factors for only two of these maturities are presented: factors for short term analyses (up to 10 years) based on the 7-year real rate (-0.4 %), and factors for long-term analyses (longer than 10 years) based on the 30-year real rate (1.1 %). As a result, these discount factors are for approximation purposes only. It is suggested that the NIST Building Life Cycle Cost (BLCC) program be used to compute the present value factors for the discount rate corresponding to the length of the study period when approximate values are not satisfactory for the project analysis. (See preface for details on obtaining this program.)

The energy price indices and corresponding present value factors published in this report are computed from energy price forecasts provided to NIST by the Department of Energy's Energy Information Administration (EIA). The EIA energy price forecast used in this report was the most recent available at the time this report was prepared. A description of the methodology used by EIA to project energy prices through 2043 is included in section B of this report. DOE has not projected escalation rates for water prices to be used in the LCC analysis of water conservation projects. Water escalation rates should be obtained from the local water utility when possible.

Federal agencies and contractors to federal agencies are encouraged to seek energy price projections from their local utility to use in place of the DOE/EIA regional projections, especially when evaluating alternative fuel types. In such cases the BLCC program can be used to calculate appropriate "modified uniform present value" (UPV*) factors for use in the LCC analysis of federal energy conservation or renewable resource projects. Otherwise, 10 CFR 436 requires the use of the DOE energy price forecasts when conducting LCC analyses of such projects. The UPV* factors for energy costs presented in this report have been precalculated with the DOE forecast data. Thus the use of these UPV* factors automatically ensures that the DOE forecast data have been included in the analysis.

Most financed federal projects, such as Energy Savings Performance Contracts (ESPC), base contract payments on projected annual energy cost savings. When setting up the contract, average rates of energy price escalation over the contract term are a matter of negotiation. One consideration in setting escalation rates is the potential for future carbon pricing. Should carbon pricing legislation be enacted by the U.S. Congress, use of the EIA-based escalation rates—which do not consider carbon pricing—likely would underestimate escalation for contract payments. To assist federal agencies in considering a range of escalation rate scenarios, in 2010 FEMP introduced to the Annual Supplement a new "D" series of tables projecting potential future carbon prices and electricity-related carbon emissions rates under a range of carbon policy scenarios. Average rates of escalation may be calculated for each of these carbon policy scenarios in the Energy Escalation Rate Calculator (EERC 2.0), a BLCC companion program for financed projects. These may be considered by federal agencies for use as energy price escalation rates for contract payments.

All of the tables of discount factors contained in this report are based on real discount rates and are therefore intended for use only with economic analyses conducted in constant dollars (in which the purchasing power of the dollar is held constant). The energy price escalation rates and corresponding energy price indices for federal analyses contained in this report are also expressed in real terms. If nominal discount rates and current dollar costs (which both include inflation) are used in the LCC analyses of federal projects, choose the current-dollar-analysis option in the BLCC computer program, which uses a nominal discount rate and adds the rate of general inflation to all dollar amounts.

This report uses the term "present value" instead of "present worth" for the discount factors presented. The meaning of these two terms is considered to be identical for purposes of economic analysis. This change in terminology was made to be consistent with the terms used in the ASTM

International compilation of standards on building economics (ASTM Standards on Building Economics, 7th Edition, ASTM, West Conshohocken, PA, 2012.)

In all of the tables, the "end-of-year" discounting convention is used, that is, all factors and indices are computed to adjust future dollar amounts to present value from the end of the year in which they are expected to occur. The factors and indices in this publication which include energy price escalation rates (e.g., UPV* factors and energy price indices) were calculated using April 1, 2013 as their base date. However, these factors and indices can be used without adjustment for the LCC analysis of projects with other base dates until the release of the next revision of this publication. Adjustment of these factors and indices for differences in the month-specific base date is not generally warranted due to uncertainties in estimating future energy prices.

PART I:
TABLES FOR FEDERAL LIFE-CYCLE COST ANALYSIS

A. Single Present Value and Uniform Present Value Factors for Non-Fuel Costs

Table A-1 presents the single present value (SPV) factors for finding the present value of future non-fuel, non-annually recurring costs, such as repair and replacement costs and salvage values. The formula for finding the present value (P) of a future cost occurring in year t (C_t) is the following:

$$P = C_t \times \frac{1}{(1+d)^t} = C_t \times SPV_t,$$

where d $\quad$ = discount rate, and
$\quad$ t $\quad$ = number of time periods (years) between the present time and the time the cost is incurred.

Table A-2 presents uniform present value (UPV) factors for finding the present value of future non-fuel costs recurring annually, such as routine maintenance costs. The formula for finding the present value (P) of an annually recurring uniform cost (A) is the following:

$$P = A \times \frac{(1+d)^N - 1}{d(1+d)^N} = A \times UPV_N,$$

where d $\quad$ = discount rate, and
$\quad$ N $\quad$ = number of time periods (years) over which A recurs.

Tables A-3 (a,b,c) present modified uniform present value (UPV*) factors for finding the present value of annually recurring non-fuel costs, such as water costs, which are expected to change from year to year at a constant rate of change (or escalation rate) over the study period. The escalation rate can be positive or negative. The formula for finding the present value (P) of an annually recurring cost at base-date prices (A_0) changing at escalation rate e is the following:

$$P = A_0 \times \left(\frac{1+e}{d-e}\right)\left[1 - \left(\frac{1+e}{1+d}\right)^N\right] = A \times UPV^*_N \qquad (d \neq e)$$

or

$$P = A_0 \times N = A \times UPV^*_N \qquad (d = e),$$

where A_0 $\quad$ = annually recurring cost at base-date prices,
$\quad$ d $\quad$ = discount rate,
$\quad$ e $\quad$ = escalation rate, and
$\quad$ N $\quad$ = number of time periods (years) over which A recurs.

4

Note: if the discount rate is expressed in real terms, i.e., net of general inflation, then the escalation rate must also be expressed in real terms. If the discount rate is expressed in nominal terms, i.e., including general inflation, then the escalation rate must also be expressed in nominal terms.

In tables A-1, A-2, and A-3 (a,b,c) SPV, UPV, and UPV* factors are provided for both the DOE and the OMB Circular A-94 real discount rates current as of the date of this publication. The FEMP SPV, UPV, and UPV* factors were computed using the DOE discount rate. The FEMP factors are for finding the present value of future costs associated with federal energy and water conservation projects and renewable energy projects. The OMB SPV, UPV, and UPV* factors were computed using the OMB discount rates. The OMB factors are for finding the present value of future costs associated with most other federal projects (except those specifically exempted from OMB Circular A-94). The DOE and OMB discount rates used in computing these tables are real rates, exclusive of general price inflation. Thus the resulting discount factors are intended for use with future costs that are stated in constant dollars.

Note: We have added to table A-3a a column of UPV factors that incorporate an escalation rate of 0.5 %, the negative of the inflation rate used to calculate the DOE nominal discount rate for 2013. The UPV* factors in this column can be used to calculate present values of fixed dollar amounts when performing a constant-dollar analysis. An example might be a fixed contract payment in an ESPC project. For these fixed amounts, the assumption that in a constant-dollar analysis all cash flows change at the rate of general inflation (so that the differential escalation rate is zero) does not apply. In real terms, fixed amounts change at a differential rate equal to the negative of the inflation rate.*

Examples of How to Use the Factors:

SPV (FEMP): To compute the present value of a replacement cost expected to occur in the 8th year for an energy efficient heating system, go to Table A-1, find the 3.0 % SPV factor for year 8 (0.789), and multiply the factor by the replacement cost as of the base date.

SPV (OMB, Short-term): To compute the present value of a repair cost in the 5th year for a floor covering (non-energy related), go to Table A-1, find the -0.4 % SPV factor for year 5 (1.020), and multiply the factor by the repair cost as of the base date.

SPV (OMB, Long-term): To compute the present value of a repair cost in the 15th year for a floor covering (non-energy related), go to Table A-1, find the 1.1 % SPV factor for year 15 (0.849), and multiply the factor by the repair cost as of the base date.

UPV (FEMP): To compute the present value of an annually recurring maintenance cost for a renewable energy system over 20 years, go to Table A-2, find the 3.0 % UPV factor for 20 years (14.88), and multiply the factor by the annual maintenance cost as of the base date.

UPV (OMB, Short-term): To compute the present value of annually recurring costs of office cleaning over 10 years (for a project not primarily related to energy conservation), go to Table A-2, find the -0.4 % UPV factor for 10 years (10.22), and multiply the factor by the annual cleaning cost as of the base date.

UPV (OMB, Long-term): To compute the present value of annually recurring costs of office cleaning over 25 years (for a project not primarily related to energy conservation), go to Table A-2, find the 1.1 % UPV factor for 25 years (21.75), and multiply the factor by the annual cleaning cost as of the base date.

UPV* (all): To compute the present value of annually recurring costs of water usage which are expected to increase at 2 % faster than the rate of general inflation over 25 years, find the UPV* factor from table A-3 (a ,b, or c as appropriate) that corresponds to 2 % escalation and a 25 year study period. From table A-3a (3.0 % DOE discount rate) the corresponding UPV* factor is 22.08. Multiply this factor by the annual water cost as computed at base year prices to determine the present value of these water costs over the entire 25 years.

UPV* (negative inflation rate): To compute the present value of an annually recurring contract payment that is fixed over a contract period of 10 years, find the UPV* factor from table A-3a that corresponds to an escalation of 0.5 % and a 10-year time period. From table A-3a (3.0 % DOE discount rate) the corresponding UPV* factor is 8.76. Multiply this factor by the annual contract payment as of the base year to determine the present value of these contract payments over the entire 10-year period.

Note: UPV factors are generally applied to costs that recur annually in substantially the same amount. Examples of such costs are routine operating and maintenance costs. UPV* factors are generally applied to costs that recur annually but change from year to year at a constant escalation rate. Examples of such costs are water usage costs when they increase from year to year. These costs usually occur every year over the service period of the building life. If there is a planning/design/construction period before the service life begins, during which these annual costs are not incurred, the appropriate UPV (or UPV*) factor for the service period is the difference between the UPV (or UPV*) factor for the entire study period and the UPV (or UPV*) factor for the planning/design/construction period. For example, if the planning/ design/construction period is 3 years and the service period is 25 years, for a total study period of 28 years, the corresponding UPV factor (from Table A-2, DOE 3.0 % discount rate) is 18.76 - 2.83 = 15.93.

For further explanation and illustration of how to use these factors, see NIST Handbook 135.

Table A-1. SPV factors for finding the present value of future single costs (non-fuel)

Number of years from base date	Single Present Value (SPV) Factors		
	DOE Discount rate 3.0 %	OMB Discount Rates[a] Short term[b] -0.4 %	Long Term[c] 1.1 %
0.25	0.993	1.001	0.997
0.50	0.985	1.002	0.995
0.75	0.978	1.003	0.992
1	0.971	1.004	0.989
2	0.943	1.008	0.978
3	0.915	1.012	0.968
4	0.888	1.016	0.957
5	0.863	1.020	0.947
6	0.837	1.024	0.936
7	0.813	1.028	0.926
8	0.789	1.033	0.916
9	0.766	1.037	0.906
10	0.744	1.041	0.896
11	0.722		0.887
12	0.701		0.877
13	0.681		0.867
14	0.661		0.858
15	0.642		0.849
16	0.623		0.839
17	0.605		0.830
18	0.587		0.821
19	0.570		0.812
20	0.554		0.803
21	0.538		0.795
22	0.522		0.786
23	0.507		0.778
24	0.492		0.769
25	0.478		0.761
26	0.464		0.752
27	0.450		0.744
28	0.437		0.736
29	0.424		0.728
30	0.412		0.720

[a]OMB discount rates as of April 2013.
[b]Short-term discount rate based on OMB discount rate for 7-year study period.
[c]Long-term discount rate based on OMB discount rate for 30-year study period.

Table A-2. UPV factors for finding the present value of annually recurring uniform costs (non-fuel)

Number of years from base date	DOE Discount rate 3.0 %	OMB Discount Rates[a] Short term[b] -0.4 %	Long Term[c] 1.1 %
		Uniform Present Value (UPV) Factors	
1	0.97	1.00	0.99
2	1.91	2.01	1.97
3	2.83	3.02	2.94
4	3.72	4.04	3.89
5	4.58	5.06	4.84
6	5.42	6.08	5.78
7	6.23	7.11	6.70
8	7.02	8.15	7.62
9	7.79	9.18	8.52
10	8.53	10.22	9.42
11	9.25		10.31
12	9.95		11.18
13	10.63		12.05
14	11.30		12.91
15	11.94		13.76
16	12.56		14.60
17	13.17		15.43
18	13.75		16.25
19	14.32		17.06
20	14.88		17.87
21	15.42		18.66
22	15.94		19.45
23	16.44		20.22
24	16.94		20.99
25	17.41		21.75
26	17.88		22.51
27	18.33		23.25
28	18.76		23.99
29	19.19		24.71
30	19.60		25.43

[a]OMB discount rates as of April 2013.
[b]Short-term discount rate based on OMB discount rate for 7-year study period.
[c]Long-term discount rate based on OMB discount rate for 30-year study period.

Table A-3a. UPV* factors for finding the present value of annually recurring costs changing at a constant escalation rate, DOE discount rate.

DOE discount rate = 3.0 %

Modified Uniform Present Value (UPV*) Factors (non-fuel)

Number of years from base date	Annual rate of price change											
	-5 %	-4 %	-3 %	-2 %	-1 %	0 %	.5 %	1 %	2 %	3 %	4 %	5 %
1	0.92	0.93	0.94	0.95	0.96	0.97	0.98	0.98	0.99	1.00	1.01	1.02
2	1.77	1.80	1.83	1.86	1.89	1.91	1.93	1.94	1.97	2.00	2.03	2.06
3	2.56	2.61	2.66	2.72	2.77	2.83	2.86	2.88	2.94	3.00	3.06	3.12
4	3.28	3.37	3.45	3.54	3.63	3.72	3.76	3.81	3.90	4.00	4.10	4.20
5	3.95	4.07	4.19	4.32	4.45	4.58	4.65	4.72	4.86	5.00	5.15	5.30
6	4.56	4.72	4.89	5.06	5.24	5.42	5.51	5.61	5.80	6.00	6.21	6.42
7	5.13	5.33	5.55	5.77	5.99	6.23	6.35	6.48	6.73	7.00	7.28	7.57
8	5.66	5.90	6.16	6.44	6.72	7.02	7.17	7.33	7.66	8.00	8.36	8.73
9	6.14	6.44	6.75	7.08	7.42	7.79	7.98	8.17	8.57	9.00	9.45	9.92
10	6.58	6.93	7.30	7.68	8.09	8.53	8.76	8.99	9.48	10.00	10.55	11.13
11	7.00	7.39	7.81	8.26	8.74	9.25	9.52	9.80	10.38	11.00	11.66	12.37
12	7.37	7.82	8.30	8.81	9.36	9.95	10.27	10.59	11.27	12.00	12.78	13.63
13	7.72	8.22	8.76	9.34	9.96	10.63	10.99	11.36	12.15	13.00	13.92	14.91
14	8.05	8.59	9.19	9.83	10.54	11.30	11.70	12.12	13.02	14.00	15.06	16.22
15	8.34	8.94	9.60	10.31	11.09	11.94	12.39	12.87	13.89	15.00	16.22	17.56
16	8.62	9.27	9.98	10.76	11.62	12.56	13.07	13.60	14.74	16.00	17.39	18.92
17	8.87	9.57	10.34	11.19	12.13	13.17	13.73	14.32	15.59	17.00	18.57	20.30
18	9.10	9.85	10.68	11.60	12.62	13.75	14.37	15.02	16.43	18.00	19.76	21.72
19	9.32	10.11	11.00	11.99	13.09	14.32	15.00	15.71	17.26	19.00	20.96	23.16
20	9.52	10.36	11.30	12.36	13.54	14.88	15.61	16.38	18.08	20.00	22.17	24.63
21	9.70	10.59	11.58	12.71	13.98	15.42	16.20	17.05	18.90	21.00	23.39	26.12
22	9.87	10.80	11.85	13.04	14.40	15.94	16.79	17.69	19.70	22.00	24.63	27.65
23	10.03	11.00	12.10	13.36	14.80	16.44	17.36	18.33	20.50	23.00	25.88	29.21
24	10.17	11.18	12.34	13.66	15.18	16.94	17.91	18.96	21.29	24.00	27.14	30.79
25	10.30	11.35	12.56	13.95	15.56	17.41	18.45	19.57	22.08	25.00	28.41	32.41
26	10.42	11.51	12.77	14.23	15.91	17.88	18.98	20.17	22.85	26.00	29.70	34.06
27	10.54	11.66	12.97	14.49	16.26	18.33	19.49	20.76	23.62	27.00	31.00	35.74
28	10.64	11.80	13.16	14.73	16.59	18.76	20.00	21.34	24.38	28.00	32.31	37.45
29	10.74	11.93	13.33	14.97	16.90	19.19	20.49	21.90	25.14	29.00	33.63	39.20
30	10.82	12.05	13.50	15.20	17.21	19.60	20.97	22.46	25.88	30.00	34.97	40.98

9

Table A-3b. UPV* factors for finding the present value of annually recurring amounts changing at a constant escalation rate, OMB short-term discount rate.

OMB short-term discount rate = -0.4 %[a]

Modified Uniform Present Value (UPV*) Factors (non-fuel)

Number of years from base date	\-5 %	\-4 %	\-3 %	\-2 %	\-1 %	0 %	1 %	2 %	3 %	4 %	5 %
						Annual rate of price change					
1	0.95	0.96	0.97	0.98	0.99	1.00	1.01	1.02	1.03	1.04	1.05
2	1.86	1.89	1.92	1.95	1.98	2.01	2.04	2.07	2.10	2.13	2.17
3	2.73	2.79	2.85	2.90	2.96	3.02	3.09	3.15	3.21	3.27	3.34
4	3.56	3.65	3.75	3.84	3.94	4.04	4.14	4.25	4.35	4.46	4.57
5	4.35	4.48	4.62	4.76	4.91	5.06	5.21	5.37	5.54	5.70	5.87
6	5.10	5.29	5.48	5.67	5.87	6.08	6.30	6.53	6.76	7.00	7.25
7	5.82	6.06	6.31	6.56	6.83	7.11	7.40	7.71	8.02	8.35	8.69
8	6.50	6.80	7.12	7.44	7.79	8.15	8.52	8.92	9.33	9.77	10.22
9	7.16	7.52	7.90	8.31	8.73	9.18	9.66	10.16	10.68	11.24	11.83
10	7.78	8.21	8.67	9.16	9.67	10.22	10.81	11.43	12.08	12.78	13.52

[a]OMB discount rate as of April 2013.
Short-term discount rate based on OMB discount rate for 7-year study period.

Table A-3c. UPV* factors for finding the present value of annually recurring amounts changing at a constant escalation rate, OMB long-term discount rate.

OMB long-term discount rate = 1.1 %[a]

Modified Uniform Present Value (UPV*) Factors (non-fuel)

Number of years from base date	Annual rate of price change										
	-5 %	-4 %	-3 %	-2 %	-1 %	0 %	1 %	2 %	3 %	4 %	5 %
1	0.94	0.95	0.96	0.97	0.98	0.99	1.00	1.01	1.02	1.03	1.04
2	1.82	1.85	1.88	1.91	1.94	1.97	2.00	2.03	2.06	2.09	2.12
3	2.65	2.71	2.76	2.82	2.88	2.94	2.99	3.05	3.11	3.18	3.24
4	3.43	3.52	3.61	3.70	3.80	3.89	3.99	4.09	4.19	4.30	4.40
5	4.16	4.29	4.42	4.56	4.70	4.84	4.99	5.14	5.29	5.45	5.61
6	4.85	5.03	5.20	5.39	5.58	5.78	5.98	6.19	6.41	6.63	6.86
7	5.50	5.72	5.95	6.19	6.44	6.70	6.97	7.25	7.55	7.85	8.17
8	6.11	6.38	6.67	6.97	7.29	7.62	7.96	8.33	8.71	9.10	9.52
9	6.68	7.01	7.36	7.73	8.12	8.52	8.96	9.41	9.89	10.39	10.93
10	7.22	7.61	8.02	8.46	8.93	9.42	9.95	10.50	11.09	11.72	12.39
11	7.72	8.17	8.65	9.17	9.72	10.31	10.93	11.61	12.32	13.09	13.90
12	8.19	8.71	9.26	9.86	10.50	11.18	11.92	12.72	13.57	14.49	15.48
13	8.64	9.22	9.85	10.52	11.26	12.05	12.91	13.84	14.85	15.93	17.11
14	9.06	9.70	10.41	11.17	12.00	12.91	13.90	14.97	16.14	17.42	18.81
15	9.45	10.16	10.94	11.80	12.73	13.76	14.88	16.11	17.47	18.95	20.58
16	9.82	10.60	11.46	12.41	13.45	14.60	15.87	17.27	18.81	20.52	22.41
17	10.17	11.02	11.95	12.99	14.15	15.43	16.85	18.43	20.18	22.14	24.31
18	10.49	11.41	12.43	13.57	14.83	16.25	17.83	19.60	21.58	23.80	26.29
19	10.80	11.78	12.88	14.12	15.50	17.06	18.81	20.79	23.01	25.51	28.34
20	11.09	12.14	13.32	14.66	16.16	17.87	19.79	21.98	24.46	27.27	30.47
21	11.36	12.48	13.74	15.18	16.81	18.66	20.77	23.18	25.94	29.09	32.69
22	11.61	12.80	14.14	15.68	17.44	19.45	21.75	24.40	27.44	30.95	34.99
23	11.85	13.10	14.53	16.17	18.05	20.22	22.73	25.63	28.98	32.86	37.38
24	12.08	13.39	14.90	16.64	18.66	20.99	23.71	26.86	30.54	34.84	39.86
25	12.29	13.66	15.25	17.10	19.25	21.75	24.68	28.11	32.13	36.86	42.43
26	12.49	13.92	15.60	17.55	19.83	22.51	25.66	29.37	33.76	38.95	45.11
27	12.67	14.17	15.92	17.98	20.40	23.25	26.63	30.64	35.41	41.10	47.89
28	12.85	14.41	16.24	18.40	20.95	23.99	27.60	31.92	37.09	43.30	50.77
29	13.01	14.63	16.54	18.80	21.49	24.71	28.57	33.21	38.81	45.57	53.77
30	13.17	14.84	16.83	19.19	22.03	25.43	29.54	34.52	40.56	47.91	56.88

[a]OMB discount rate as of April 2013.
Long-term discount rate based on OMB discount rate for 30-year study period.

B.   Modified Uniform Present Value Factors for Fuel Costs

This section presents FEMP and OMB modified uniform present value (UPV*) discount factors for calculating the present value of energy usage for federal projects. Factors are provided for the four major Census regions and for the overall United States. The factors are modified in the sense that they incorporate energy price escalation rates based on future energy prices projected by DOE for the years 2013 to 2043. There are two sets of UPV* tables: the "Ba" tables present FEMP UPV* factors based on the DOE discount rate (3.0 % real), and the "Bb" tables present OMB UPV* factors based on two OMB discount rates (-0.4 % real for short-term study periods of 1 to 10 years, 1.1 % real for long-term study periods of 11 to 30 years). The underlying energy price indices for the years 2013 to 2043, on which these UPV* calculations are based, are shown in tables Ca-1 through Ca-5. The corresponding average energy price escalation rates for selected time intervals between 2013 and 2043 are shown in tables Cb-1 through Cb-5.

Energy Price Projections. The FEMP and OMB UPV* factors incorporate energy price escalation rates computed from future energy prices projected by the Energy Information Administration (EIA) of the U.S. Department of Energy. Energy prices through 2040 were generated by EIA using the National Energy Modeling System (NEMS) and published in the *Annual Energy Outlook 2013* (AEO2013). At the request of FEMP, EIA extended its price projections from 2040 to 2043 based on a combination of the NEMS model and extrapolations from the AEO2013 projections.

NEMS is an energy market model designed to project the impacts of alternative energy policies or assumptions on U.S. energy markets. NEMS produces projections of the U.S. energy future, given current laws and policies and other key assumptions, including macroeconomic indicators from Data Resources, Inc., the production policy of the Organization of Petroleum Exporting Countries, the size of the economically recoverable resource base for fossil fuels, and the rate of development and penetration of new technologies. NEMS balances energy supply and demands with modules representing primary fuel supply, end-use demand for four sectors, and conversion of energy by refineries and electricity generators. Macroeconomic and international oil modules reflect the impacts of energy prices, production, and consumption on world oil markets and the economy.

The EIA energy price projections presented in this report, like those of other forecasts, are dependent on the data, methodologies, and specific assumptions used in their development. Many of the assumptions concerning the future cannot be known with any degree of certainty. Thus, the projections are not statements of what will happen, but what might happen given the particular assumptions and methodologies used. Although EIA has endeavored to make these forecasts as objective, reliable, and useful as possible, these projections should serve as an adjunct to, not a substitute for, the analytical process. The AEO2013 was prepared by EIA as required under statute by federal legislation. The price projections to 2043 were prepared in accordance with a Service Request from the Federal Energy Management Program.

Note: Section 441 of the Energy Independence and Security Act of 2007 (EISA) extends from 25 years to 40 years the maximum service period for conducting FEMP life-cycle cost analyses. To account for the legislated change, the BLCC program now incorporates unofficial projections of future energy prices beyond 2043 to accommodate FEMP service periods of up to 40 years. The projections are based on simple extrapolations of 2043 growth rates and are not reported here because they are not endorsed by EIA. BLCC users should exercise caution when interpreting energy cost savings beyond 30 years and do sensitivity analyses to test different out-year assumptions.

UPV* Calculation Method. The formula for finding the present value (P) of future energy costs or savings is the following:

$$P = A_0 x \sum_{t=1}^{N} \frac{I_{(2013+t)}}{(1+d)^t} = A_0 x UPV_N^*$$

where A_0 = annual cost of energy as of the base date (April 1, 2013);
      t   = index used to designate the year of energy usage;
      N   = number of periods, e.g., years, over which energy costs or savings accrue;
      $I_{(2013+t)}$ = projected average fuel price index[1] given in Tables Ca-1 through Ca-5
                   for the year 2013+t (where I_{2013} = 1.00); and
      d   = the real discount rate.

This formula is based on end-of-year energy prices and end-of-year discounting. Note that annual energy costs as of the base date of the LCC analysis (A_0, to be supplied by the analyst) should reflect the current energy price schedule as of that date, which may not be the same as the energy price itself on that date.[2] That is, the annual energy cost should reflect summer-winter rate differences, time-of-use rates, block rates considerations, and demand charges (as appropriate) anticipated to be in effect that year. If energy and demand costs are calculated separately (as is sometimes done for electricity), the UPV* factor should be applied to both costs.

The data in the tables that follow are reported for the four Census regions and the U.S. average. Figure B-1 presents a map showing the states corresponding to the four Census regions. The Census regions do not include American Samoa, Canal Zone, Guam, Puerto Rico, Trust Territory of the Pacific Islands, or the Virgin Islands. Analysts of federal projects in these areas should use data that are "reasonable under the circumstances," and may refer to the tables with U.S. average data for guidance.

---

[1] For greater precision, the UPV* factors reported in the Ba and Bb tables were computed using the unrounded form of the indices given in Tables Ca-1 through Ca-5.

[2] While the UPV* factors provided in this publication were computed using energy price indices that correspond to energy prices as of April 1 in the current and future years, the analyst is encouraged to use for determining A_0 the energy prices prevailing as of the base date of the LCC analysis for the project evaluated.

Figure B-1

Source: U.S. Census Bureau

14

B.1. FEMP Modified Uniform Present Value Factors

The FEMP Modified Uniform Present Value (FEMP UPV*) factors presented in the "Ba" tables, based on the current DOE discount rate (3.0 %), are for calculating the present value of energy costs or savings accruing over 1 to 30 years and are to be used in life-cycle cost analyses of federal energy conservation and renewable energy projects. These factors may be applied to projects with or without planning/design/construction periods, as shown below.

These factors apply only to annual energy usage or energy savings that are assumed to be the same each year over the service period. The BLCC computer program can compute the present value of energy usage and savings that are not the same in each year.

Examples of How to Use the FEMP UPV* Factors:

FEMP UPV*, no planning/design/construction period: To compute the present value of heating with natural gas over 25 years in a federal office building in New Mexico, go to Table Ba-4, find the FEMP UPV* factor for commercial natural gas for 25 years (22.63), and multiply this factor by the annual heating cost at base-date natural gas prices.

FEMP UPV*, with planning/design/construction period: To compute a present value factor for a service period following a planning/design/construction period (1) find the FEMP UPV* factor for the combined length of the planning/design/construction period and the service period, and (2) subtract from (1) the FEMP UPV* factor for the planning/design/ construction period alone. The difference is the FEMP UPV* factor for the years over which energy costs or savings actually accrue. For example, suppose a new federal office building in New York is being evaluated with several energy conserving design options. It is expected to have a planning/design/construction period of 5 years, after which it will be occupied for 25 years. To compute the present value of natural gas costs over 25 years of occupancy, go to Table Ba-1 and find the FEMP UPV* factors for commercial natural gas for 5 years (4.61) and for 30 years (22.92). The difference (18.31) is the FEMP UPV* factor for natural gas costs over 25 years, beginning 5 years after the base date. Multiply 18.31 by the annual natural gas cost at base date prices (not occupancy-date prices) to calculate the present value of natural gas costs over the entire 25-year occupancy period.

15

Table Ba-1. FEMP UPV* Discount Factors adjusted for fuel price escalation, by end-use sector and fuel type.

Discount rate = 3.0 % (DOE)

Census Region 1 (Connecticut, Maine, Massachusetts, New Hampshire, New Jersey, New York, Pennsylvania, Rhode Island, Vermont)

N	RESIDENTIAL				COMMERCIAL					INDUSTRIAL					TRANSPORT	N
	Elec	Dist	LPG	NtGas	Elec	Dist	Resid	NtGas	Coal	Elec	Dist	Resid	NtGas	Coal	GasIn	
1	0.97	0.95	0.94	0.96	0.95	0.91	0.73	0.96	0.97	0.95	0.89	0.76	0.97	0.96	0.94	1
2	1.90	1.87	1.83	1.88	1.86	1.77	1.37	1.88	1.92	1.85	1.74	1.44	1.91	1.92	1.84	2
3	2.81	2.78	2.69	2.79	2.74	2.63	2.01	2.79	2.85	2.72	2.57	2.12	2.86	2.86	2.72	3
4	3.69	3.67	3.55	3.70	3.60	3.47	2.63	3.70	3.77	3.58	3.39	2.78	3.81	3.79	3.57	4
5	4.54	4.55	4.40	4.60	4.44	4.30	3.26	4.61	4.70	4.41	4.21	3.45	4.77	4.70	4.41	5
6	5.36	5.41	5.25	5.49	5.26	5.13	3.88	5.51	5.61	5.22	5.02	4.10	5.72	5.60	5.23	6
7	6.15	6.27	6.08	6.37	6.06	5.94	4.50	6.39	6.51	6.01	5.81	4.76	6.65	6.48	6.06	7
8	6.91	7.11	6.90	7.23	6.84	6.74	5.11	7.25	7.38	6.79	6.60	5.41	7.56	7.34	6.87	8
9	7.65	7.94	7.70	8.08	7.59	7.54	5.72	8.10	8.24	7.54	7.38	6.05	8.46	8.17	7.67	9
10	8.37	8.75	8.49	8.91	8.32	8.32	6.33	8.93	9.08	8.29	8.15	6.69	9.35	9.00	8.45	10
11	9.08	9.56	9.27	9.73	9.03	9.09	6.93	9.75	9.90	9.02	8.91	7.33	10.23	9.80	9.22	11
12	9.78	10.35	10.03	10.54	9.72	9.86	7.53	10.55	10.71	9.72	9.66	7.96	11.08	10.58	9.97	12
13	10.47	11.13	10.78	11.33	10.36	10.61	8.12	11.34	11.49	10.39	10.40	8.59	11.94	11.34	10.70	13
14	11.14	11.90	11.51	12.11	10.99	11.35	8.72	12.11	12.26	11.05	11.13	9.21	12.77	12.09	11.41	14
15	11.81	12.66	12.22	12.87	11.61	12.08	9.31	12.87	13.01	11.70	11.84	9.83	13.59	12.82	12.11	15
16	12.46	13.41	12.92	13.61	12.22	12.80	9.89	13.61	13.74	12.34	12.55	10.45	14.39	13.53	12.80	16
17	13.10	14.15	13.60	14.34	12.81	13.51	10.48	14.34	14.46	12.96	13.24	11.06	15.18	14.22	13.48	17
18	13.71	14.87	14.27	15.06	13.39	14.20	11.05	15.05	15.16	13.57	13.92	11.66	15.95	14.89	14.14	18
19	14.32	15.58	14.92	15.77	13.95	14.88	11.62	15.75	15.85	14.17	14.58	12.26	16.71	15.55	14.79	19
20	14.91	16.28	15.55	16.46	14.51	15.56	12.18	16.43	16.52	14.76	15.24	12.85	17.46	16.19	15.43	20
21	15.48	16.98	16.17	17.15	15.04	16.23	12.74	17.11	17.17	15.32	15.89	13.44	18.20	16.81	16.06	21
22	16.04	17.66	16.78	17.82	15.57	16.88	13.30	17.78	17.81	15.88	16.54	14.03	18.93	17.42	16.68	22
23	16.59	18.33	17.38	18.49	16.09	17.54	13.86	18.44	18.44	16.43	17.18	14.61	19.67	18.02	17.30	23
24	17.13	19.00	17.97	19.16	16.60	18.18	14.41	19.10	19.06	16.98	17.81	15.19	20.40	18.61	17.91	24
25	17.66	19.66	18.54	19.81	17.10	18.82	14.96	19.76	19.66	17.52	18.43	15.76	21.13	19.18	18.51	25
26	18.18	20.30	19.11	20.46	17.60	19.44	15.51	20.41	20.25	18.07	19.04	16.34	21.86	19.75	19.10	26
27	18.69	20.94	19.66	21.10	18.09	20.06	16.05	21.05	20.83	18.60	19.65	16.90	22.58	20.29	19.69	27
28	19.18	21.57	20.20	21.73	18.57	20.68	16.59	21.68	21.39	19.12	20.24	17.47	23.30	20.83	20.27	28
29	19.67	22.19	20.73	22.35	19.04	21.28	17.13	22.30	21.94	19.63	20.83	18.03	24.01	21.36	20.85	29
30	20.14	22.81	21.25	22.96	19.50	21.88	17.67	22.92	22.48	20.13	21.42	18.59	24.72	21.87	21.41	30

Table Ba-2. FEMP UPV* Discount Factors adjusted for fuel price escalation, by end-use sector and fuel type.

Discount Rate = 3.0 % (DOE)

Census Region 2 (Illinois, Indiana, Iowa, Kansas, Michigan, Minnesota,
Missouri, Nebraska, North Dakota, Ohio, South Dakota, Wisconsin)

N	RESIDENTIAL				COMMERCIAL					INDUSTRIAL					TRANSPORT	N
	Elec	Dist	LPG	NtGas	Elec	Dist	Resid	NtGas	Coal	Elec	Dist	Resid	NtGas	Coal	Gasln	
1	0.98	0.90	0.94	0.95	0.97	0.84	1.01	0.96	0.96	0.96	0.85	1.05	0.99	0.97	0.93	1
2	1.93	1.74	1.83	1.84	1.90	1.63	2.00	1.86	1.90	1.88	1.64	2.07	1.96	1.92	1.82	2
3	2.88	2.58	2.70	2.73	2.84	2.40	2.96	2.77	2.83	2.80	2.42	3.08	2.96	2.86	2.69	3
4	3.82	3.40	3.55	3.64	3.76	3.16	3.90	3.69	3.73	3.71	3.20	4.06	3.98	3.78	3.53	4
5	4.74	4.22	4.41	4.55	4.67	3.92	4.84	4.63	4.62	4.61	3.96	5.04	5.04	4.68	4.36	5
6	5.63	5.02	5.25	5.47	5.56	4.67	5.76	5.57	5.49	5.48	4.72	6.00	6.09	5.57	5.18	6
7	6.50	5.81	6.08	6.37	6.41	5.41	6.67	6.50	6.35	6.32	5.47	6.95	7.14	6.44	6.00	7
8	7.33	6.59	6.90	7.26	7.24	6.15	7.57	7.41	7.19	7.13	6.21	7.88	8.17	7.28	6.80	8
9	8.15	7.37	7.71	8.14	8.04	6.87	8.45	8.31	8.01	7.93	6.95	8.81	9.20	8.11	7.59	9
10	8.94	8.13	8.50	9.01	8.83	7.59	9.33	9.20	8.81	8.71	7.68	9.72	10.22	8.92	8.37	10
11	9.72	8.88	9.28	9.87	9.60	8.30	10.20	10.08	9.59	9.47	8.39	10.62	11.23	9.72	9.12	11
12	10.47	9.62	10.04	10.71	10.34	9.00	11.06	10.95	10.35	10.21	9.10	11.52	12.22	10.50	9.86	12
13	11.21	10.36	10.79	11.53	11.06	9.69	11.90	11.79	11.09	10.93	9.80	12.40	13.19	11.26	10.58	13
14	11.93	11.08	11.52	12.34	11.76	10.38	12.74	12.61	11.82	11.64	10.49	13.27	14.14	12.01	11.29	14
15	12.62	11.80	12.24	13.13	12.44	11.05	13.57	13.42	12.53	12.32	11.16	14.13	15.07	12.75	11.98	15
16	13.30	12.50	12.93	13.91	13.10	11.71	14.38	14.22	13.23	12.99	11.83	14.98	15.99	13.46	12.66	16
17	13.95	13.20	13.62	14.67	13.74	12.37	15.19	15.00	13.91	13.64	12.48	15.82	16.89	14.17	13.32	17
18	14.59	13.88	14.29	15.43	14.36	13.01	15.99	15.76	14.57	14.27	13.12	16.65	17.78	14.86	13.98	18
19	15.21	14.56	14.94	16.16	14.97	13.65	16.78	16.51	15.22	14.89	13.76	17.47	18.66	15.54	14.62	19
20	15.81	15.22	15.58	16.89	15.56	14.27	17.56	17.26	15.85	15.50	14.38	18.28	19.52	16.20	15.25	20
21	16.39	15.88	16.20	17.62	16.14	14.89	18.33	17.99	16.47	16.09	15.00	19.07	20.39	16.84	15.88	21
22	16.96	16.53	16.81	18.34	16.70	15.51	19.09	18.73	17.08	16.67	15.61	19.87	21.26	17.48	16.49	22
23	17.52	17.18	17.41	19.07	17.26	16.11	19.85	19.47	17.68	17.25	16.22	20.65	22.15	18.10	17.10	23
24	18.07	17.81	18.00	19.80	17.81	16.72	20.59	20.22	18.26	17.82	16.81	21.43	23.05	18.71	17.71	24
25	18.61	18.44	18.58	20.53	18.35	17.31	21.33	20.97	18.83	18.39	17.41	22.19	23.95	19.31	18.30	25
26	19.13	19.06	19.14	21.25	18.87	17.90	22.06	21.71	19.39	18.94	17.99	22.95	24.86	19.90	18.89	26
27	19.63	19.68	19.70	21.96	19.39	18.48	22.78	22.44	19.93	19.49	18.57	23.70	25.75	20.47	19.47	27
28	20.13	20.28	20.24	22.66	19.89	19.05	23.50	23.16	20.47	20.02	19.13	24.44	26.64	21.04	20.05	28
29	20.61	20.88	20.77	23.36	20.38	19.62	24.21	23.88	20.99	20.54	19.70	25.18	27.53	21.59	20.62	29
30	21.07	21.47	21.30	24.04	20.86	20.18	24.91	24.59	21.50	21.05	20.26	25.91	28.41	22.13	21.18	30

17

Table Ba-3. FEMP UPV* Discount Factors adjusted for fuel price escalation, by end-use sector and fuel type.

Discount Rate = 3.0 % (DOE)

Census Region 3 (Alabama, Arkansas, Delaware, District of Columbia, Florida, Georgia, Kentucky, Louisiana, Maryland, Mississippi, North Carolina, Oklahoma, South Carolina, Tennessee, Texas, Virginia, West Virginia)

N	RESIDENTIAL				COMMERCIAL					INDUSTRIAL					TRANSPORT	N
	Elec	Dist	LPG	NtGas	Elec	Dist	Resid	NtGas	Coal	Elec	Dist	Resid	NtGas	Coal	GasIn	
1	0.98	0.94	0.94	0.95	0.97	0.86	0.82	0.96	0.97	0.97	0.88	0.85	1.03	0.97	0.93	1
2	1.94	1.85	1.83	1.85	1.92	1.67	1.58	1.87	1.91	1.91	1.70	1.64	2.05	1.92	1.82	2
3	2.89	2.74	2.70	2.76	2.86	2.46	2.32	2.78	2.84	2.84	2.52	2.42	3.13	2.85	2.68	3
4	3.82	3.63	3.55	3.66	3.77	3.25	3.06	3.70	3.73	3.76	3.32	3.18	4.24	3.76	3.52	4
5	4.73	4.49	4.40	4.55	4.67	4.03	3.78	4.61	4.59	4.67	4.12	3.94	5.37	4.65	4.34	5
6	5.61	5.35	5.25	5.44	5.54	4.80	4.50	5.52	5.43	5.54	4.91	4.69	6.49	5.51	5.16	6
7	6.47	6.19	6.08	6.31	6.39	5.56	5.21	6.41	6.23	6.40	5.69	5.43	7.60	6.35	5.97	7
8	7.30	7.02	6.90	7.17	7.21	6.31	5.92	7.29	7.01	7.23	6.46	6.17	8.70	7.17	6.77	8
9	8.12	7.84	7.70	8.01	8.02	7.05	6.62	8.16	7.77	8.05	7.23	6.90	9.80	7.97	7.56	9
10	8.91	8.65	8.50	8.85	8.80	7.79	7.31	9.03	8.50	8.85	7.99	7.62	10.91	8.75	8.33	10
11	9.67	9.45	9.27	9.68	9.56	8.51	8.00	9.88	9.22	9.63	8.73	8.33	12.03	9.51	9.09	11
12	10.42	10.23	10.03	10.49	10.29	9.23	8.68	10.72	9.93	10.39	9.47	9.04	13.12	10.25	9.83	12
13	11.15	11.01	10.78	11.29	11.00	9.94	9.35	11.54	10.61	11.13	10.19	9.74	14.21	10.98	10.56	13
14	11.86	11.77	11.51	12.07	11.69	10.64	10.02	12.35	11.28	11.85	10.91	10.43	15.28	11.68	11.27	14
15	12.55	12.52	12.22	12.84	12.37	11.32	10.68	13.14	11.94	12.56	11.61	11.12	16.34	12.37	11.96	15
16	13.23	13.26	12.92	13.59	13.03	12.00	11.34	13.92	12.58	13.26	12.30	11.80	17.38	13.05	12.64	16
17	13.89	13.99	13.60	14.33	13.67	12.67	11.99	14.69	13.20	13.94	12.98	12.48	18.41	13.71	13.31	17
18	14.53	14.71	14.27	15.06	14.29	13.33	12.62	15.44	13.82	14.60	13.65	13.14	19.43	14.35	13.97	18
19	15.15	15.42	14.92	15.78	14.91	13.97	13.26	16.17	14.42	15.25	14.30	13.80	20.43	14.98	14.62	19
20	15.76	16.12	15.55	16.48	15.50	14.61	13.89	16.90	15.00	15.89	14.95	14.45	21.42	15.60	15.25	20
21	16.35	16.80	16.17	17.18	16.08	15.24	14.50	17.62	15.57	16.52	15.59	15.10	22.42	16.20	15.88	21
22	16.93	17.48	16.78	17.87	16.66	15.87	15.12	18.33	16.13	17.14	16.22	15.74	23.43	16.79	16.50	22
23	17.50	18.15	17.38	18.57	17.23	16.49	15.73	19.05	16.68	17.75	16.85	16.37	24.46	17.37	17.11	23
24	18.06	18.82	17.97	19.26	17.79	17.10	16.34	19.76	17.22	18.36	17.47	17.00	25.52	17.94	17.72	24
25	18.61	19.47	18.54	19.94	18.34	17.70	16.94	20.48	17.75	18.97	18.08	17.63	26.59	18.49	18.32	25
26	19.15	20.11	19.10	20.62	18.89	18.30	17.54	21.19	18.27	19.57	18.69	18.25	27.65	19.04	18.91	26
27	19.68	20.75	19.66	21.29	19.42	18.89	18.13	21.88	18.77	20.16	19.29	18.86	28.72	19.57	19.49	27
28	20.20	21.38	20.20	21.94	19.95	19.47	18.71	22.57	19.26	20.74	19.88	19.47	29.78	20.09	20.07	28
29	20.70	22.00	20.73	22.59	20.46	20.05	19.30	23.25	19.75	21.31	20.46	20.07	30.84	20.60	20.65	29
30	21.19	22.61	21.25	23.23	20.96	20.62	19.88	23.93	20.22	21.86	21.04	20.67	31.90	21.09	21.21	30

18

Table Ba-4. FEMP UPV* Discount Factors adjusted for fuel price escalation, by end-use sector and fuel type.

Discount Rate = 3.0 % (DOE)

Census Region 4 (Alaska, Arizona, California, Colorado, Hawaii,
Idaho, Montana, Nevada, New Mexico, Oregon, Utah, Washington, Wyoming)

N	RESIDENTIAL				COMMERCIAL					INDUSTRIAL					TRANSPORT	N
	Elec	Dist	LPG	NtGas	Elec	Dist	Resid	NtGas	Coal	Elec	Dist	Resid	NtGas	Coal	Gasln	
1	0.97	0.89	0.94	0.97	0.97	0.85	0.72	0.98	0.98	0.96	0.87	0.75	1.01	0.97	0.94	1
2	1.92	1.73	1.83	1.94	1.92	1.65	1.36	1.93	1.95	1.89	1.70	1.42	2.01	1.93	1.84	2
3	2.84	2.56	2.69	2.92	2.84	2.43	1.98	2.91	2.91	2.80	2.51	2.08	3.04	2.86	2.71	3
4	3.73	3.38	3.55	3.93	3.74	3.21	2.59	3.89	3.84	3.69	3.32	2.72	4.11	3.76	3.56	4
5	4.61	4.19	4.40	4.94	4.62	3.99	3.20	4.90	4.76	4.57	4.13	3.37	5.20	4.63	4.41	5
6	5.44	4.99	5.24	5.96	5.46	4.76	3.81	5.91	5.65	5.41	4.92	4.01	6.31	5.47	5.24	6
7	6.26	5.78	6.07	6.96	6.28	5.51	4.40	6.92	6.53	6.24	5.71	4.65	7.41	6.29	6.07	7
8	7.05	6.56	6.89	7.96	7.09	6.26	5.00	7.91	7.38	7.04	6.49	5.28	8.51	7.10	6.89	8
9	7.81	7.32	7.70	8.95	7.86	7.00	5.58	8.90	8.22	7.81	7.26	5.91	9.60	7.88	7.70	9
10	8.54	8.08	8.48	9.93	8.60	7.73	6.17	9.88	9.05	8.56	8.02	6.54	10.69	8.65	8.49	10
11	9.25	8.83	9.26	10.89	9.32	8.45	6.75	10.85	9.86	9.29	8.77	7.17	11.76	9.41	9.26	11
12	9.95	9.56	10.02	11.83	10.02	9.17	7.33	11.79	10.65	10.01	9.51	7.80	12.79	10.15	10.02	12
13	10.63	10.29	10.76	12.75	10.70	9.87	7.91	12.71	11.43	10.71	10.23	8.43	13.79	10.88	10.76	13
14	11.28	11.01	11.49	13.64	11.36	10.57	8.48	13.61	12.20	11.39	10.95	9.05	14.76	11.60	11.48	14
15	11.92	11.71	12.20	14.52	12.00	11.25	9.04	14.49	12.96	12.05	11.66	9.66	15.70	12.30	12.19	15
16	12.54	12.41	12.89	15.38	12.61	11.92	9.60	15.35	13.71	12.69	12.35	10.27	16.63	12.99	12.88	16
17	13.15	13.10	13.57	16.22	13.21	12.59	10.16	16.19	14.44	13.32	13.04	10.88	17.54	13.67	13.57	17
18	13.74	13.78	14.23	17.05	13.80	13.24	10.71	17.02	15.17	13.94	13.71	11.48	18.44	14.34	14.24	18
19	14.33	14.44	14.88	17.87	14.38	13.89	11.26	17.84	15.89	14.55	14.37	12.08	19.32	15.00	14.90	19
20	14.90	15.10	15.51	18.68	14.93	14.52	11.81	18.64	16.60	15.14	15.02	12.68	20.20	15.65	15.54	20
21	15.45	15.75	16.13	19.47	15.48	15.15	12.35	19.44	17.29	15.72	15.66	13.26	21.08	16.29	16.18	21
22	15.99	16.39	16.74	20.27	16.01	15.77	12.88	20.24	17.98	16.30	16.30	13.85	21.96	16.92	16.81	22
23	16.53	17.02	17.33	21.06	16.54	16.38	13.41	21.04	18.66	16.87	16.93	14.43	22.84	17.54	17.44	23
24	17.06	17.65	17.91	21.85	17.06	16.99	13.94	21.83	19.33	17.43	17.55	15.01	23.74	18.15	18.05	24
25	17.57	18.27	18.48	22.64	17.57	17.59	14.46	22.63	19.99	17.99	18.17	15.59	24.65	18.74	18.66	25
26	18.08	18.88	19.04	23.42	18.08	18.18	14.98	23.43	20.64	18.54	18.77	16.17	25.55	19.33	19.26	26
27	18.58	19.48	19.59	24.20	18.57	18.77	15.50	24.21	21.28	19.09	19.37	16.74	26.45	19.90	19.86	27
28	19.07	20.07	20.12	24.96	19.05	19.34	16.01	24.99	21.91	19.62	19.96	17.30	27.34	20.46	20.45	28
29	19.54	20.65	20.65	25.71	19.53	19.92	16.52	25.75	22.52	20.15	20.55	17.87	28.22	21.01	21.03	29
30	20.01	21.23	21.16	26.45	19.98	20.48	17.03	26.51	23.12	20.67	21.12	18.43	29.10	21.55	21.60	30

19

Table Ba-5. FEMP UPV* Discount Factors adjusted for fuel price escalation, by end-use sector and fuel type.

Discount Rate = 3.0 % (DOE)

United States Average

N	RESIDENTIAL				COMMERCIAL					INDUSTRIAL					TRANSPORT	N
	Elec	Dist	LPG	NtGas	Elec	Dist	Resid	NtGas	Coal	Elec	Dist	Resid	NtGas	Coal	GasIn	
1	0.98	0.95	0.94	0.96	0.97	0.88	0.74	0.96	0.97	0.96	0.87	0.84	1.02	0.97	0.94	1
2	1.93	1.86	1.83	1.87	1.90	1.70	1.39	1.88	1.92	1.89	1.69	1.62	2.01	1.92	1.83	2
3	2.86	2.75	2.69	2.79	2.82	2.52	2.03	2.80	2.85	2.81	2.49	2.39	3.05	2.86	2.69	3
4	3.78	3.64	3.55	3.72	3.73	3.33	2.67	3.73	3.76	3.71	3.29	3.14	4.12	3.78	3.54	4
5	4.67	4.51	4.40	4.65	4.61	4.12	3.30	4.67	4.65	4.60	4.08	3.89	5.20	4.67	4.37	5
6	5.53	5.37	5.24	5.57	5.47	4.91	3.93	5.60	5.53	5.46	4.87	4.64	6.29	5.54	5.19	6
7	6.37	6.22	6.08	6.49	6.30	5.69	4.56	6.52	6.38	6.30	5.64	5.37	7.36	6.40	6.01	7
8	7.18	7.05	6.90	7.38	7.11	6.46	5.18	7.42	7.22	7.12	6.40	6.10	8.42	7.23	6.82	8
9	7.97	7.87	7.70	8.27	7.90	7.22	5.80	8.32	8.03	7.91	7.16	6.82	9.48	8.04	7.62	9
10	8.73	8.68	8.49	9.15	8.66	7.97	6.41	9.21	8.83	8.69	7.91	7.53	10.54	8.84	8.40	10
11	9.48	9.48	9.27	10.02	9.40	8.72	7.02	10.08	9.61	9.45	8.65	8.24	11.59	9.62	9.16	11
12	10.21	10.27	10.03	10.86	10.11	9.45	7.63	10.93	10.37	10.19	9.38	8.94	12.62	10.38	9.90	12
13	10.92	11.05	10.78	11.70	10.81	10.17	8.23	11.77	11.11	10.91	10.10	9.64	13.64	11.13	10.63	13
14	11.61	11.81	11.51	12.51	11.48	10.88	8.83	12.59	11.84	11.61	10.81	10.32	14.63	11.85	11.34	14
15	12.28	12.57	12.22	13.31	12.13	11.59	9.42	13.40	12.55	12.30	11.50	11.01	15.61	12.57	12.04	15
16	12.94	13.31	12.92	14.10	12.77	12.28	10.01	14.18	13.25	12.97	12.19	11.68	16.57	13.26	12.73	16
17	13.58	14.04	13.60	14.87	13.39	12.96	10.60	14.96	13.94	13.62	12.86	12.35	17.52	13.95	13.40	17
18	14.20	14.76	14.27	15.62	13.99	13.63	11.18	15.72	14.61	14.26	13.52	13.01	18.45	14.62	14.06	18
19	14.81	15.47	14.92	16.37	14.58	14.29	11.75	16.46	15.27	14.89	14.18	13.67	19.37	15.28	14.71	19
20	15.40	16.17	15.56	17.10	15.16	14.94	12.33	17.19	15.91	15.50	14.82	14.31	20.29	15.92	15.35	20
21	15.97	16.86	16.18	17.83	15.72	15.58	12.89	17.92	16.55	16.09	15.45	14.95	21.20	16.54	15.98	21
22	16.53	17.54	16.79	18.55	16.27	16.22	13.45	18.65	17.16	16.69	16.08	15.59	22.12	17.16	16.60	22
23	17.09	18.21	17.39	19.28	16.82	16.84	14.02	19.37	17.77	17.28	16.70	16.22	23.06	17.77	17.22	23
24	17.63	18.87	17.98	20.00	17.35	17.47	14.58	20.10	18.37	17.86	17.32	16.85	24.01	18.36	17.83	24
25	18.16	19.53	18.55	20.71	17.88	18.08	15.13	20.82	18.95	18.44	17.93	17.47	24.97	18.94	18.43	25
26	18.68	20.17	19.12	21.42	18.40	18.69	15.68	21.54	19.53	19.01	18.52	18.09	25.92	19.51	19.02	26
27	19.19	20.81	19.67	22.12	18.91	19.29	16.23	22.24	20.09	19.57	19.12	18.70	26.87	20.07	19.61	27
28	19.69	21.44	20.21	22.81	19.41	19.88	16.77	22.94	20.64	20.12	19.70	19.30	27.82	20.61	20.19	28
29	20.17	22.06	20.74	23.49	19.89	20.46	17.31	23.64	21.18	20.66	20.28	19.91	28.76	21.15	20.77	29
30	20.65	22.67	21.26	24.16	20.37	21.04	17.85	24.32	21.70	21.19	20.85	20.51	29.70	21.67	21.33	30

20

B.2. OMB Modified Uniform Present Value Factors

The OMB Modified Uniform Present Value (OMB UPV*) factors presented in the "Bb" tables, based on the current OMB discount rates (-0.4 % short term and 1.1 % long term), are for calculating the present value of energy costs accruing over 1 to 30 years when conducting a life-cycle cost analysis of a federal project not explicitly related to energy or water conservation or renewable resources. These factors apply only to annual energy usage that is assumed to be the same each year over the service period. The BLCC computer program can compute the present value of energy usage and savings that are not the same in each year.

Examples of How to Use the OMB UPV* Factors:

OMB UPV* (OMB discount rate): To compute the present value over 30 years of electricity costs associated with the occupancy of a federal office building in Ohio (where energy conservation is not a specific consideration in the LCC analysis), go to Table Bb-2, find the OMB UPV* factor for commercial electricity for 30 years (27.22), and multiply this factor by the annual electricity cost in base-date dollars.

Note: Because the discount rate used to calculate the Bb tables (OMB discount rate) is usually different for years 1 to 10 than for years 11 to 30, these factors cannot be used with a planning/ design/construction period as shown above for the Ba tables (DOE discount rate). Use the BLCC computer program for this purpose. For further explanation of the use of UPV* factors, see NIST Handbook 135.

Table Bb-1. OMB UPV* Discount Factors adjusted for fuel price escalation, by end-use sector and fuel type.

Discount Rate = -0.4 % (years 1 to 10) and 1.1 % (years 11 to 30), (OMB Circular A-94[a])

Census Region 1 (Connecticut, Maine, Massachusetts, New Hampshire, New Jersey, New York, Pennsylvania, Rhode Island, Vermont)

N	RESIDENTIAL				COMMERCIAL					INDUSTRIAL					TRANSPORT	N
	Elec	Dist	LPG	NtGas	Elec	Dist	Resid	NtGas	Coal	Elec	Dist	Resid	NtGas	Coal	GasIn	
1	1.00	0.98	0.97	0.99	0.98	0.94	0.75	1.00	1.00	0.98	0.92	0.79	1.00	1.00	0.97	1
2	2.00	1.97	1.93	1.97	1.95	1.87	1.44	1.98	2.02	1.94	1.83	1.52	2.01	2.02	1.94	2
3	3.00	2.97	2.88	2.98	2.92	2.81	2.14	2.99	3.05	2.91	2.75	2.26	3.06	3.06	2.90	3
4	4.01	3.99	3.86	4.02	3.91	3.77	2.86	4.03	4.10	3.89	3.69	3.02	4.15	4.12	3.88	4
5	5.01	5.03	4.87	5.09	4.90	4.76	3.60	5.10	5.19	4.87	4.66	3.81	5.28	5.20	4.87	5
6	6.01	6.09	5.90	6.18	5.91	5.77	4.36	6.20	6.31	5.87	5.64	4.61	6.44	6.30	5.88	6
7	7.01	7.17	6.95	7.29	6.93	6.79	5.14	7.31	7.45	6.87	6.65	5.44	7.62	7.41	6.92	7
8	8.01	8.27	8.02	8.42	7.94	7.84	5.94	8.44	8.59	7.88	7.68	6.29	8.81	8.53	7.98	8
9	9.01	9.39	9.11	9.56	8.95	8.92	6.77	9.58	9.76	8.90	8.74	7.16	10.03	9.67	9.07	9
10	10.03	10.53	10.22	10.73	9.98	10.01	7.62	10.75	10.93	9.95	9.82	8.05	11.27	10.82	10.16	10
11	10.12	10.67	10.35	10.87	10.07	10.16	7.75	10.89	11.06	10.06	9.96	8.19	11.43	10.94	10.29	11
12	10.99	11.67	11.30	11.88	10.92	11.11	8.49	11.89	12.07	10.93	10.90	8.98	12.51	11.91	11.23	12
13	11.86	12.66	12.25	12.89	11.75	12.07	9.25	12.90	13.07	11.79	11.84	9.78	13.59	12.89	12.16	13
14	12.74	13.66	13.20	13.90	12.56	13.03	10.02	13.90	14.06	12.65	12.78	10.59	14.68	13.86	13.09	14
15	13.62	14.67	14.14	14.90	13.38	14.00	10.80	14.91	15.05	13.50	13.73	11.41	15.76	14.82	14.01	15
16	14.50	15.67	15.08	15.91	14.20	14.97	11.59	15.90	16.04	14.36	14.67	12.24	16.84	15.77	14.94	16
17	15.37	16.68	16.02	16.91	15.01	15.94	12.39	16.90	17.03	15.21	15.62	13.08	17.92	16.72	15.87	17
18	16.23	17.70	16.95	17.92	15.82	16.91	13.19	17.89	18.01	16.06	16.57	13.92	19.00	17.66	16.79	18
19	17.09	18.71	17.88	18.92	16.62	17.88	14.00	18.89	18.99	16.92	17.52	14.77	20.08	18.60	17.72	19
20	17.95	19.73	18.80	19.93	17.43	18.86	14.82	19.88	19.96	17.77	18.48	15.63	21.16	19.53	18.65	20
21	18.79	20.75	19.72	20.94	18.22	19.85	15.65	20.88	20.93	18.61	19.44	16.50	22.26	20.45	19.58	21
22	19.64	21.78	20.64	21.96	19.01	20.84	16.49	21.89	21.89	19.44	20.41	17.38	23.36	21.37	20.52	22
23	20.48	22.81	21.56	22.99	19.81	21.84	17.35	22.91	22.86	20.30	21.39	18.28	24.49	22.29	21.47	23
24	21.33	23.86	22.47	24.02	20.60	22.85	18.21	23.94	23.82	21.16	22.37	19.19	25.64	23.21	22.42	24
25	22.16	24.90	23.39	25.07	21.40	23.86	19.09	24.99	24.78	22.02	23.36	20.10	26.81	24.12	23.37	25
26	23.01	25.95	24.30	26.12	22.21	24.88	19.97	26.04	25.73	22.90	24.36	21.03	27.99	25.03	24.34	26
27	23.85	27.01	25.22	27.18	23.02	25.90	20.87	27.09	26.69	23.78	25.36	21.97	29.18	25.94	25.31	27
28	24.69	28.07	26.13	28.24	23.83	26.93	21.78	28.16	27.64	24.66	26.37	22.92	30.38	26.84	26.29	28
29	25.52	29.14	27.04	29.31	24.64	27.97	22.71	29.23	28.59	25.54	27.38	23.89	31.61	27.75	27.27	29
30	26.34	30.21	27.95	30.38	25.44	29.02	23.65	30.31	29.53	26.41	28.40	24.87	32.85	28.64	28.26	30

22

[a]OMB discount rate as of April 2013.

Table Bb-2. OMB UPV* Discount Factors adjusted for fuel price escalation, by end-use sector and fuel type.

Discount Rate = -0.4 % (years 1 to 10) and 1.1 % (years 11 to 30), (OMB Circular A-94[a])

Census Region 2 (Illinois, Indiana, Iowa, Kansas, Michigan, Minnesota, Missouri, Nebraska, North Dakota, Ohio, South Dakota, Wisconsin)

N	RESIDENTIAL				COMMERCIAL					INDUSTRIAL					TRANSPORT	N
---	Elec	Dist	LPG	NtGas	Elec	Dist	Resid	NtGas	Coal	Elec	Dist	Resid	NtGas	Coal	Gasln	
1	1.01	0.93	0.97	0.98	1.00	0.87	1.05	0.99	0.99	0.99	0.88	1.08	1.03	1.00	0.97	1
2	2.03	1.83	1.93	1.93	2.00	1.71	2.10	1.95	2.00	1.98	1.73	2.18	2.06	2.02	1.92	2
3	3.08	2.76	2.88	2.92	3.03	2.56	3.16	2.96	3.02	2.99	2.59	3.29	3.16	3.06	2.87	3
4	4.16	3.70	3.86	3.95	4.09	3.44	4.24	4.02	4.05	4.03	3.48	4.42	4.34	4.12	3.84	4
5	5.25	4.66	4.87	5.04	5.17	4.34	5.35	5.13	5.11	5.09	4.38	5.57	5.58	5.18	4.82	5
6	6.34	5.64	5.90	6.16	6.25	5.25	6.47	6.28	6.18	6.16	5.31	6.74	6.88	6.26	5.82	6
7	7.43	6.65	6.96	7.30	7.33	6.19	7.62	7.45	7.26	7.22	6.26	7.94	8.20	7.36	6.85	7
8	8.52	7.67	8.03	8.46	8.41	7.15	8.80	8.64	8.36	8.29	7.23	9.17	9.55	8.46	7.90	8
9	9.63	8.71	9.12	9.65	9.50	8.13	10.00	9.86	9.46	9.36	8.22	10.42	10.94	9.59	8.98	9
10	10.74	9.78	10.23	10.87	10.60	9.14	11.23	11.11	10.58	10.45	9.24	11.70	12.37	10.72	10.06	10
11	10.84	9.92	10.36	11.04	10.70	9.27	11.39	11.29	10.70	10.56	9.38	11.87	12.59	10.85	10.18	11
12	11.78	10.85	11.32	12.09	11.64	10.15	12.46	12.36	11.65	11.49	10.26	12.98	13.82	11.82	11.11	12
13	12.72	11.78	12.27	13.14	12.55	11.03	13.54	13.44	12.60	12.41	11.15	14.10	15.06	12.80	12.02	13
14	13.65	12.72	13.22	14.19	13.46	11.92	14.63	14.51	13.54	13.32	12.04	15.23	16.29	13.77	12.94	14
15	14.57	13.67	14.16	15.24	14.36	12.81	15.72	15.58	14.48	14.23	12.94	16.37	17.53	14.74	13.85	15
16	15.48	14.62	15.11	16.29	15.24	13.70	16.82	16.65	15.42	15.12	13.83	17.52	18.77	15.70	14.77	16
17	16.38	15.57	16.04	17.33	16.12	14.60	17.94	17.72	16.35	16.02	14.73	18.67	20.00	16.67	15.68	17
18	17.27	16.53	16.98	18.38	16.99	15.50	19.05	18.79	17.28	16.90	15.63	19.83	21.24	17.64	16.60	18
19	18.15	17.49	17.91	19.44	17.86	16.40	20.17	19.86	18.21	17.78	16.53	21.00	22.49	18.60	17.51	19
20	19.02	18.46	18.83	20.50	18.71	17.31	21.30	20.93	19.12	18.66	17.44	22.17	23.75	19.56	18.43	20
21	19.89	19.43	19.76	21.57	19.57	18.23	22.44	22.03	20.04	19.54	18.35	23.35	25.03	20.52	19.36	21
22	20.75	20.41	20.68	22.66	20.42	19.15	23.59	23.14	20.96	20.42	19.27	24.55	26.35	21.47	20.28	22
23	21.61	21.40	21.60	23.77	21.28	20.09	24.75	24.28	21.87	21.31	20.20	25.75	27.71	22.43	21.22	23
24	22.46	22.40	22.52	24.91	22.13	21.03	25.92	25.44	22.78	22.20	21.14	26.96	29.11	23.38	22.16	24
25	23.31	23.40	23.44	26.07	22.99	21.98	27.09	26.63	23.69	23.10	22.08	28.18	30.55	24.34	23.11	25
26	24.16	24.41	24.36	27.24	23.85	22.93	28.28	27.83	24.59	24.00	23.02	29.41	32.02	25.29	24.06	26
27	24.99	25.42	25.27	28.42	24.70	23.89	29.47	29.04	25.50	24.90	23.98	30.65	33.50	26.24	25.03	27
28	25.82	26.44	26.18	29.60	25.54	24.85	30.68	30.26	26.39	25.80	24.94	31.90	35.00	27.19	26.00	28
29	26.65	27.46	27.10	30.79	26.39	25.83	31.90	31.50	27.29	26.69	25.90	33.17	36.52	28.13	26.98	29
30	27.46	28.49	28.01	31.99	27.22	26.81	33.13	32.74	28.18	27.58	26.88	34.44	38.06	29.08	27.96	30

[a]OMB discount rate as of April 2013.

23

Table Bb-3. OMB UPV* Discount Factors adjusted for fuel price escalation, by end-use sector and fuel type.

Discount Rate = -0.4 % (years 1 to 10) and 1.1 % (years 11 to 30), (OMB Circular A-94[a])

Census Region 3 (Alabama, Arkansas, Delaware, District of Columbia, Florida, Georgia, Kentucky, Louisiana, Maryland, Mississippi, North Carolina, Oklahoma, South Carolina, Tennessee, Texas, Virginia, West Virginia)

N	RESIDENTIAL				COMMERCIAL					INDUSTRIAL					TRANSPORT	N
	Elec	Dist	LPG	NtGas	Elec	Dist	Resid	NtGas	Coal	Elec	Dist	Resid	NtGas	Coal	GasIn	
1	1.02	0.97	0.97	0.98	1.01	0.89	0.85	0.99	1.00	1.00	0.91	0.88	1.07	1.00	0.96	1
2	2.04	1.95	1.93	1.95	2.02	1.75	1.66	1.96	2.01	2.01	1.79	1.73	2.16	2.02	1.91	2
3	3.09	2.93	2.88	2.95	3.05	2.63	2.48	2.97	3.03	3.04	2.69	2.58	3.35	3.05	2.86	3
4	4.16	3.94	3.86	3.98	4.10	3.53	3.32	4.02	4.05	4.09	3.61	3.46	4.62	4.09	3.82	4
5	5.23	4.97	4.87	5.04	5.17	4.45	4.18	5.11	5.08	5.16	4.56	4.35	5.96	5.14	4.80	5
6	6.31	6.02	5.90	6.12	6.23	5.39	5.06	6.22	6.10	6.23	5.53	5.27	7.33	6.20	5.79	6
7	7.39	7.08	6.95	7.22	7.30	6.35	5.96	7.34	7.11	7.31	6.51	6.21	8.73	7.26	6.82	7
8	8.48	8.17	8.02	8.34	8.38	7.34	6.88	8.49	8.13	8.40	7.52	7.17	10.16	8.33	7.87	8
9	9.58	9.28	9.11	9.49	9.47	8.34	7.83	9.67	9.16	9.51	8.56	8.16	11.66	9.41	8.94	9
10	10.69	10.41	10.22	10.66	10.56	9.37	8.80	10.88	10.19	10.63	9.61	9.17	13.22	10.50	10.02	10
11	10.79	10.55	10.35	10.82	10.65	9.51	8.94	11.05	10.27	10.74	9.76	9.31	13.49	10.60	10.15	11
12	11.72	11.53	11.31	11.83	11.57	10.41	9.78	12.10	11.15	11.69	10.68	10.19	14.86	11.53	11.08	12
13	12.65	12.52	12.26	12.85	12.48	11.31	10.64	13.15	12.03	12.63	11.60	11.08	16.24	12.45	12.00	13
14	13.57	13.51	13.20	13.86	13.37	12.22	11.51	14.20	12.89	13.57	12.53	11.98	17.63	13.37	12.92	14
15	14.49	14.51	14.15	14.88	14.27	13.13	12.38	15.24	13.76	14.51	13.45	12.90	19.03	14.28	13.84	15
16	15.40	15.51	15.08	15.89	15.16	14.04	13.27	16.29	14.62	15.45	14.39	13.81	20.44	15.19	14.76	16
17	16.30	16.51	16.02	16.91	16.04	14.96	14.16	17.34	15.48	16.38	15.32	14.74	21.85	16.10	15.67	17
18	17.20	17.51	16.95	17.93	16.91	15.87	15.05	18.39	16.34	17.31	16.25	15.67	23.27	17.00	16.59	18
19	18.08	18.52	17.88	18.95	17.78	16.80	15.95	19.44	17.19	18.24	17.19	16.61	24.70	17.90	17.51	19
20	18.96	19.53	18.80	19.97	18.65	17.72	16.86	20.49	18.04	19.16	18.13	17.55	26.13	18.79	18.44	20
21	19.84	20.55	19.72	21.00	19.51	18.66	17.78	21.55	18.89	20.09	19.07	18.51	27.61	19.68	19.37	21
22	20.71	21.57	20.64	22.05	20.38	19.60	18.71	22.63	19.73	21.02	20.03	19.47	29.13	20.57	20.30	22
23	21.59	22.60	21.56	23.11	21.25	20.55	19.65	23.73	20.58	21.97	20.99	20.45	30.72	21.46	21.24	23
24	22.47	23.63	22.47	24.19	22.12	21.50	20.60	24.85	21.42	22.93	21.96	21.43	32.37	22.35	22.19	24
25	23.35	24.67	23.39	25.28	23.00	22.46	21.55	25.99	22.26	23.89	22.94	22.42	34.07	23.23	23.14	25
26	24.22	25.72	24.30	26.38	23.89	23.43	22.52	27.14	23.09	24.86	23.92	23.43	35.80	24.11	24.10	26
27	25.10	26.77	25.21	27.48	24.77	24.41	23.50	28.29	23.93	25.84	24.90	24.44	37.56	24.99	25.07	27
28	25.96	27.82	26.12	28.59	25.66	25.39	24.48	29.45	24.76	26.81	25.90	25.47	39.35	25.87	26.05	28
29	26.83	28.89	27.03	29.70	26.53	26.38	25.49	30.62	25.59	27.79	26.90	26.50	41.17	26.74	27.03	29
30	27.69	29.96	27.94	30.82	27.40	27.38	26.50	31.79	26.42	28.76	27.91	27.55	43.03	27.61	28.02	30

[a]OMB discount rate as of April 2013.

24

Table Bb-4. OMB UPV* Discount Factors adjusted for fuel price escalation, by end-use sector and fuel type.

Discount Rate = -0.4 % (years 1 to 10) and 1.1 % (years 11 to 30), (OMB Circular A-94[a])

Census Region 4 (Alaska, Arizona, California, Colorado, Hawaii, Idaho, Montana, Nevada, New Mexico, Oregon, Utah, Washington, Wyoming)

N	RESIDENTIAL				COMMERCIAL					INDUSTRIAL					TRANSPORT	N
	Elec	Dist	LPG	NtGas	Elec	Dist	Resid	NtGas	Coal	Elec	Dist	Resid	NtGas	Coal	Gasln	
1	1.01	0.92	0.97	1.00	1.01	0.88	0.75	1.01	1.01	1.00	0.90	0.78	1.05	1.01	0.97	1
2	2.02	1.82	1.93	2.04	2.02	1.73	1.43	2.03	2.05	1.99	1.79	1.49	2.11	2.03	1.93	2
3	3.03	2.74	2.88	3.13	3.04	2.60	2.11	3.11	3.11	3.00	2.69	2.22	3.25	3.06	2.89	3
4	4.06	3.67	3.86	4.28	4.07	3.49	2.82	4.24	4.18	4.02	3.61	2.96	4.47	4.09	3.87	4
5	5.09	4.63	4.86	5.48	5.10	4.41	3.54	5.43	5.26	5.05	4.57	3.72	5.77	5.12	4.87	5
6	6.11	5.61	5.89	6.72	6.14	5.35	4.28	6.67	6.35	6.08	5.54	4.51	7.13	6.15	5.89	6
7	7.14	6.61	6.95	7.99	7.18	6.30	5.03	7.94	7.46	7.12	6.53	5.32	8.52	7.19	6.94	7
8	8.17	7.63	8.01	9.30	8.23	7.28	5.81	9.24	8.58	8.17	7.55	6.14	9.95	8.24	8.02	8
9	9.20	8.66	9.10	10.63	9.27	8.28	6.60	10.58	9.72	9.22	8.59	6.99	11.43	9.30	9.11	9
10	10.23	9.72	10.21	12.00	10.31	9.31	7.42	11.95	10.87	10.27	9.65	7.87	12.95	10.38	10.21	10
11	10.31	9.86	10.34	12.20	10.38	9.45	7.54	12.15	11.00	10.36	9.80	8.02	13.19	10.49	10.35	11
12	11.18	10.78	11.29	13.37	11.26	10.34	8.27	13.33	11.99	11.26	10.72	8.80	14.48	11.42	11.29	12
13	12.04	11.71	12.23	14.54	12.13	11.24	9.00	14.51	12.98	12.14	11.65	9.60	15.76	12.35	12.23	13
14	12.89	12.64	13.18	15.70	12.99	12.14	9.74	15.67	13.99	13.03	12.58	10.40	17.01	13.28	13.17	14
15	13.73	13.57	14.12	16.86	13.82	13.04	10.49	16.83	14.99	13.90	13.51	11.22	18.26	14.21	14.11	15
16	14.57	14.51	15.05	18.02	14.65	13.95	11.24	17.99	15.99	14.77	14.45	12.04	19.51	15.14	15.04	16
17	15.40	15.46	15.98	19.18	15.48	14.86	12.00	19.15	17.01	15.63	15.39	12.87	20.76	16.07	15.98	17
18	16.23	16.40	16.91	20.34	16.30	15.77	12.78	20.31	18.02	16.50	16.33	13.72	22.01	17.01	16.92	18
19	17.07	17.35	17.83	21.50	17.11	16.69	13.56	21.47	19.04	17.36	17.27	14.57	23.27	17.95	17.85	19
20	17.89	18.31	18.75	22.67	17.93	17.61	14.35	22.64	20.07	18.22	18.21	15.43	24.55	18.89	18.80	20
21	18.71	19.26	19.66	23.85	18.73	18.54	15.15	23.82	21.11	19.08	19.16	16.30	25.84	19.84	19.74	21
22	19.53	20.23	20.58	25.05	19.54	19.47	15.96	25.02	22.14	19.95	20.12	17.19	27.17	20.79	20.69	22
23	20.35	21.20	21.49	26.27	20.35	20.42	16.77	26.24	23.18	20.82	21.09	18.08	28.53	21.74	21.65	23
24	21.17	22.18	22.40	27.50	21.16	21.37	17.59	27.49	24.23	21.70	22.06	18.98	29.93	22.69	22.61	24
25	22.00	23.16	23.30	28.76	21.97	22.32	18.42	28.77	25.28	22.59	23.04	19.91	31.37	23.63	23.58	25
26	22.82	24.15	24.21	30.03	22.79	23.28	19.27	30.05	26.33	23.49	24.03	20.84	32.84	24.58	24.55	26
27	23.64	25.14	25.11	31.31	23.61	24.25	20.12	31.35	27.39	24.39	25.01	21.78	34.32	25.53	25.54	27
28	24.47	26.14	26.02	32.59	24.42	25.22	20.99	32.66	28.45	25.29	26.01	22.73	35.82	26.47	26.53	28
29	25.28	27.15	26.92	33.88	25.23	26.20	21.86	33.97	29.50	26.20	27.01	23.70	37.34	27.41	27.52	29
30	26.09	28.16	27.82	35.17	26.03	27.19	22.75	35.29	30.54	27.10	28.02	24.69	38.88	28.35	28.52	30

[a]OMB discount rate as of April 2013.

Table Bb-5. OMB UPV* Discount Factors adjusted for fuel price escalation, by end-use sector and fuel type.

Discount Rate = -0.4 % (years 1 to 10) and 1.1 % (years 11 to 30), (OMB Circular A-94[a])

United States Average

N	RESIDENTIAL				COMMERCIAL					INDUSTRIAL					TRANSPORT	N
	Elec	Dist	LPG	NtGas	Elec	Dist	Resid	NtGas	Coal	Elec	Dist	Resid	NtGas	Coal	GasIn	
1	1.01	0.98	0.97	0.99	1.00	0.91	0.76	1.00	1.00	1.00	0.90	0.87	1.05	1.00	0.97	1
2	2.03	1.95	1.93	1.97	2.00	1.79	1.46	1.98	2.01	1.99	1.77	1.71	2.11	2.02	1.92	2
3	3.06	2.95	2.88	2.99	3.02	2.69	2.17	3.00	3.05	3.00	2.66	2.55	3.26	3.06	2.88	3
4	4.11	3.96	3.86	4.05	4.05	3.62	2.90	4.06	4.09	4.04	3.58	3.42	4.48	4.11	3.84	4
5	5.16	4.99	4.87	5.15	5.10	4.56	3.65	5.17	5.14	5.09	4.51	4.30	5.77	5.16	4.83	5
6	6.22	6.04	5.90	6.28	6.15	5.52	4.42	6.31	6.21	6.14	5.47	5.21	7.10	6.23	5.84	6
7	7.28	7.11	6.95	7.43	7.20	6.51	5.21	7.47	7.29	7.20	6.45	6.14	8.45	7.31	6.87	7
8	8.34	8.20	8.02	8.61	8.26	7.51	6.02	8.65	8.39	8.27	7.45	7.09	9.84	8.40	7.93	8
9	9.40	9.31	9.11	9.81	9.32	8.55	6.86	9.86	9.49	9.35	8.48	8.07	11.27	9.50	9.00	9
10	10.48	10.45	10.22	11.04	10.39	9.60	7.72	11.10	10.60	10.43	9.53	9.07	12.75	10.61	10.10	10
11	10.57	10.59	10.35	11.20	10.48	9.74	7.85	11.27	10.72	10.54	9.67	9.21	12.99	10.73	10.22	11
12	11.47	11.58	11.30	12.26	11.37	10.65	8.60	12.34	11.67	11.46	10.58	10.08	14.28	11.68	11.16	12
13	12.38	12.57	12.25	13.32	12.25	11.58	9.37	13.41	12.62	12.38	11.50	10.97	15.57	12.63	12.08	13
14	13.28	13.56	13.20	14.38	13.12	12.50	10.15	14.47	13.56	13.29	12.41	11.86	16.86	13.58	13.01	14
15	14.17	14.56	14.15	15.44	13.99	13.43	10.93	15.54	14.51	14.20	13.33	12.77	18.16	14.52	13.93	15
16	15.05	15.56	15.09	16.50	14.85	14.36	11.73	16.60	15.45	15.10	14.26	13.67	19.46	15.46	14.85	16
17	15.93	16.56	16.02	17.55	15.70	15.29	12.54	17.66	16.39	16.00	15.18	14.60	20.75	16.40	15.78	17
18	16.80	17.57	16.96	18.61	16.54	16.23	13.35	18.72	17.33	16.89	16.11	15.52	22.06	17.34	16.70	18
19	17.66	18.58	17.88	19.67	17.38	17.17	14.16	19.78	18.27	17.78	17.03	16.45	23.37	18.27	17.63	19
20	18.52	19.59	18.81	20.74	18.22	18.11	14.99	20.85	19.20	18.67	17.97	17.39	24.70	19.20	18.56	20
21	19.37	20.61	19.73	21.82	19.05	19.06	15.83	21.92	20.13	19.56	18.90	18.33	26.05	20.13	19.49	21
22	20.22	21.64	20.65	22.91	19.88	20.02	16.68	23.01	21.07	20.46	19.85	19.29	27.43	21.06	20.42	22
23	21.07	22.67	21.57	24.01	20.71	20.98	17.54	24.13	22.00	21.36	20.81	20.26	28.87	21.99	21.37	23
24	21.91	23.70	22.49	25.14	21.55	21.96	18.41	25.26	22.93	22.27	21.77	21.24	30.35	22.92	22.32	24
25	22.76	24.74	23.40	26.28	22.39	22.94	19.29	26.41	23.86	23.19	22.74	22.23	31.88	23.84	23.28	25
26	23.61	25.79	24.32	27.43	23.23	23.92	20.19	27.57	24.79	24.11	23.71	23.23	33.43	24.76	24.24	26
27	24.45	26.84	25.23	28.59	24.08	24.91	21.09	28.74	25.72	25.04	24.69	24.24	35.00	25.69	25.21	27
28	25.29	27.90	26.14	29.75	24.92	25.91	22.01	29.92	26.65	25.96	25.67	25.26	36.60	26.60	26.19	28
29	26.12	28.96	27.05	30.92	25.75	26.91	22.94	31.11	27.57	26.89	26.66	26.30	38.22	27.52	27.18	29
30	26.94	30.03	27.96	32.09	26.58	27.92	23.89	32.31	28.49	27.81	27.66	27.34	39.87	28.43	28.17	30

26

[a]OMB discount rate as of April 2013.

C.   Projected Average Fuel Price Indices and Escalation Rates (Real)

Tables Ca-1 through Ca-5 present projected fuel price indices for the four Census regions and for the United States. These indices, when multiplied by annual energy costs computed at base-date prices (i.e., as of April 1, 2013), provide estimates of future-year costs (also as of April 1) in constant base-date dollars. Constant-dollar cost estimates are needed when discounting is performed with a real discount rate (i.e., a rate that does not include general price inflation).

These indices were used in the calculation of the UPV* factors for energy prices in the Ba and Bb tables in this publication. While they are based on April 1 energy prices to maintain consistency in the computation of these UPV* factors, the level of precision implied here is not required for most LCC analyses. That is, the analyst need not calibrate base-year energy prices precisely to April 1, 2013 levels to use these indices (or the corresponding UPV* factors); instead, the analyst should use current price levels as of the base-date of the LCC analysis, regardless of the time of the year that the study is undertaken.

Example of How to Use the Indices:

To estimate the price of industrial coal in 2016 in Connecticut (in constant 2013 dollars), go to Table Ca-1, find the year 2016 index for industrial coal (1.03), and multiply by the price for industrial coal in Connecticut in 2013.

For further explanation of how to use these tables, see NIST Handbook 135.

Tables Cb-1 through Cb-5 present the projected average fuel price escalation rates (percentage change compounded annually) for selected periods from 2013 to 2043 for the four Census regions and for the overall United States. Note that these are real rates exclusive of general price inflation. Their use results in prices expressed in *constant* dollars.

The average fuel escalation rates consolidate the information provided by the indices in the Ca tables so that trends in projected price changes can be seen at a glance. They are provided primarily to accommodate computer programs (such as BLCC) which require price escalation rates as inputs.

Unless there is a compelling reason to use escalation rates, it is recommended that you use the indices in the Ca tables when you need estimates of future-year energy prices, since the indices include year-to-year information rather than averages over a number of years.

Example of How to Use the Escalation Rates:

To estimate the unit price of residential natural gas at the end of 2023 (p_{23}) in Wyoming using the DOE energy price escalation rates, go to Table Cb-4 and find the 2013 to 2018 and the 2018 to 2023 escalation rates for residential natural gas (3.3 % for 5 years and 2.3 % for 5 years, respectively). Enter these values and the unit price of residential natural gas in Wyoming in 2013 (p_{13}) into the following formula. Then solve for the 2023 energy price (stated in 2013 dollars):

$$p_y = p_0 \times \prod_{i=1}^{N} (1+e_i)^{k_i}$$

$$
\begin{aligned}
p_{23} &= p_{13} \times (1+e_1)^{k_1} \times (1+e_2)^{k_2} \\
&= p_{13} \times (1+0.033)^5 \times (1+0.023)^5 \\
&= p_{13} \times 1.176 \times 1.12 \\
&= p_{13} \times 1.317
\end{aligned}
$$

where p_y = price at end of year y;

p_0 = unit price at base date;

e_i = annual compound escalation rate for period i from the Cb tables (in decimal form); and

k_i = number of years over which escalation rate e_i occurs.

Note that the compounded escalation rate factor (1.317) corresponds to the fuel price index in region 4, residential natural gas, for the year 2022 in table Ca-4 (1.32).

The data in the Ca and Cb tables on the following pages are reported for the four Census regions and the U.S. average. Figure B-1 on page 13 presents a map showing the states corresponding to the four Census regions. The Census regions do not include American Samoa, Canal Zone, Guam, Puerto Rico, Trust Territory of the Pacific Islands, or the Virgin Islands. Analysts of federal projects in these areas should use data which are "reasonable under the circumstances," and may refer to the tables with U.S. average data for guidance.

Table Ca-1. Projected fuel price indices (excluding general inflation), by end-use sector and fuel type.

Census Region 1 (Connecticut, Maine, Massachusetts, New Hampshire, New Jersey, New York, Pennsylvania, Rhode Island, Vermont)

Projected April 1 Fuel Price Indices (April 1, 2013 = 1.00)

Sector and Fuel	2014	2015	2016	2017	2018	2019	2020	2021	2022	2023	2024	2025	2026	2027	2028
Residential															
Electricity	0.99	0.99	0.99	0.99	0.99	0.98	0.97	0.97	0.96	0.97	0.98	1.00	1.01	1.02	1.04
Distillate Oil	0.98	0.98	0.99	1.00	1.02	1.03	1.05	1.06	1.08	1.10	1.11	1.13	1.15	1.17	1.18
LPG	0.97	0.94	0.94	0.96	0.99	1.01	1.02	1.04	1.05	1.06	1.07	1.08	1.09	1.10	1.11
Natural Gas	0.99	0.97	1.00	1.02	1.05	1.07	1.08	1.09	1.11	1.12	1.13	1.14	1.17	1.17	1.19
Commercial															
Electricity	0.98	0.96	0.96	0.97	0.98	0.99	0.98	0.98	0.98	0.98	0.99	0.98	0.95	0.95	0.97
Distillate Oil	0.94	0.92	0.93	0.95	0.97	0.98	1.00	1.02	1.04	1.05	1.07	1.09	1.10	1.12	1.14
Residual Oil	0.75	0.68	0.69	0.71	0.73	0.74	0.76	0.78	0.80	0.82	0.84	0.85	0.87	0.90	0.92
Natural Gas	0.99	0.97	1.00	1.02	1.05	1.07	1.08	1.09	1.11	1.12	1.13	1.14	1.16	1.17	1.18
Coal	1.00	1.01	1.02	1.03	1.07	1.09	1.10	1.11	1.12	1.13	1.14	1.14	1.15	1.16	1.17
Industrial															
Electricity	0.98	0.96	0.96	0.96	0.97	0.97	0.97	0.98	0.99	1.01	1.01	1.00	0.99	1.00	1.01
Distillate Oil	0.92	0.90	0.91	0.93	0.95	0.96	0.98	1.00	1.02	1.04	1.05	1.07	1.08	1.10	1.11
Residual Oil	0.78	0.73	0.74	0.75	0.77	0.78	0.80	0.82	0.84	0.86	0.88	0.90	0.92	0.94	0.97
Natural Gas	1.00	1.00	1.04	1.07	1.11	1.13	1.15	1.15	1.17	1.20	1.21	1.22	1.25	1.26	1.27
Coal	0.99	1.01	1.03	1.04	1.06	1.07	1.08	1.09	1.09	1.10	1.11	1.11	1.12	1.13	1.13
Transportation															
Motor Gasoline	0.97	0.96	0.95	0.96	0.97	0.99	1.01	1.03	1.04	1.05	1.06	1.07	1.07	1.08	1.09

Table Ca-1, continued. Projected fuel price indices (excluding general inflation), by end-use sector and fuel type.

Census Region 1 (Connecticut, Maine, Massachusetts, New Hampshire, New Jersey, New York, Pennsylvania, Rhode Island, Vermont)

Projected April 1 Fuel Price Indices (April 1, 2013 = 1.00)

Sector and Fuel	2029	2030	2031	2032	2033	2034	2035	2036	2037	2038	2039	2040	2041	2042	2043
Residential															
Electricity	1.04	1.05	1.05	1.06	1.07	1.06	1.07	1.09	1.10	1.10	1.12	1.13	1.13	1.14	1.15
Distillate Oil	1.20	1.22	1.23	1.25	1.27	1.29	1.31	1.33	1.36	1.37	1.40	1.42	1.44	1.47	1.49
LPG	1.12	1.13	1.13	1.14	1.15	1.16	1.17	1.18	1.19	1.20	1.21	1.22	1.24	1.25	1.26
Natural Gas	1.20	1.21	1.22	1.24	1.25	1.27	1.30	1.32	1.35	1.38	1.40	1.42	1.44	1.46	1.48
Commercial															
Electricity	0.97	0.98	0.98	0.99	1.00	1.00	1.01	1.02	1.03	1.05	1.08	1.09	1.10	1.11	1.11
Distillate Oil	1.15	1.17	1.18	1.20	1.22	1.24	1.26	1.29	1.31	1.33	1.35	1.38	1.40	1.43	1.45
Residual Oil	0.94	0.96	0.97	1.00	1.02	1.04	1.07	1.10	1.13	1.15	1.18	1.21	1.24	1.27	1.30
Natural Gas	1.19	1.20	1.21	1.22	1.24	1.26	1.28	1.31	1.34	1.38	1.40	1.42	1.45	1.47	1.50
Coal	1.18	1.19	1.20	1.20	1.21	1.22	1.23	1.24	1.25	1.26	1.27	1.28	1.29	1.30	1.31
Industrial															
Electricity	1.02	1.03	1.04	1.05	1.07	1.05	1.06	1.10	1.12	1.13	1.17	1.18	1.19	1.20	1.21
Distillate Oil	1.13	1.14	1.16	1.17	1.19	1.21	1.23	1.26	1.28	1.30	1.32	1.35	1.37	1.39	1.42
Residual Oil	0.99	1.01	1.02	1.05	1.07	1.09	1.12	1.15	1.18	1.20	1.23	1.26	1.29	1.33	1.36
Natural Gas	1.29	1.30	1.32	1.33	1.35	1.38	1.41	1.45	1.49	1.54	1.57	1.60	1.64	1.68	1.72
Coal	1.14	1.14	1.15	1.15	1.16	1.16	1.17	1.18	1.19	1.20	1.21	1.22	1.23	1.24	1.25
Transportation															
Motor Gasoline	1.10	1.12	1.13	1.14	1.16	1.17	1.19	1.21	1.24	1.26	1.28	1.31	1.33	1.35	1.38

30

Table Ca-2. Projected fuel price indices (excluding general inflation), by end-use sector and fuel type.

Census Region 2 (Illinois, Indiana, Iowa, Kansas, Michigan, Minnesota,
Missouri, Nebraska, North Dakota, Ohio, South Dakota, Wisconsin)

Projected April 1 Fuel Price Indices (April 1, 2013 = 1.00)

Sector and Fuel	2014	2015	2016	2017	2018	2019	2020	2021	2022	2023	2024	2025	2026	2027	2028
Residential															
Electricity	1.00	1.01	1.04	1.06	1.07	1.06	1.06	1.06	1.06	1.07	1.07	1.08	1.08	1.08	1.08
Distillate Oil	0.92	0.90	0.91	0.93	0.94	0.96	0.97	0.99	1.01	1.02	1.04	1.06	1.08	1.10	1.11
LPG	0.97	0.95	0.94	0.97	0.99	1.01	1.02	1.04	1.05	1.07	1.08	1.09	1.10	1.11	1.11
Natural Gas	0.97	0.94	0.98	1.02	1.06	1.09	1.11	1.13	1.15	1.17	1.19	1.20	1.21	1.22	1.24
Commercial															
Electricity	1.00	1.00	1.02	1.04	1.05	1.05	1.05	1.05	1.05	1.06	1.06	1.06	1.06	1.06	1.06
Distillate Oil	0.87	0.83	0.84	0.86	0.88	0.90	0.91	0.93	0.95	0.97	0.98	1.00	1.02	1.03	1.05
Residual Oil	1.04	1.04	1.05	1.06	1.08	1.10	1.12	1.14	1.16	1.18	1.20	1.22	1.24	1.27	1.29
Natural Gas	0.99	0.96	0.99	1.04	1.09	1.12	1.14	1.15	1.18	1.20	1.22	1.23	1.24	1.25	1.26
Coal	0.99	1.00	1.01	1.02	1.03	1.05	1.06	1.06	1.07	1.07	1.08	1.09	1.09	1.10	1.11
Industrial															
Electricity	0.99	0.98	1.00	1.03	1.04	1.04	1.03	1.03	1.04	1.04	1.05	1.06	1.06	1.06	1.07
Distillate Oil	0.87	0.84	0.85	0.87	0.89	0.91	0.92	0.94	0.96	0.98	0.99	1.01	1.02	1.04	1.05
Residual Oil	1.08	1.09	1.10	1.11	1.13	1.15	1.17	1.19	1.21	1.23	1.25	1.27	1.29	1.32	1.34
Natural Gas	1.02	1.02	1.09	1.16	1.22	1.26	1.28	1.31	1.34	1.37	1.40	1.41	1.43	1.44	1.46
Coal	1.00	1.01	1.03	1.04	1.04	1.06	1.06	1.07	1.08	1.09	1.10	1.11	1.12	1.13	1.14
Transportation															
Motor Gasoline	0.96	0.94	0.94	0.95	0.96	0.98	1.00	1.02	1.03	1.04	1.05	1.05	1.06	1.07	1.08

31

Table Ca-2, continued. Projected fuel price indices (excluding general inflation), by end-use sector and fuel type.

Census Region 2 (Illinois, Indiana, Iowa, Kansas, Michigan, Minnesota, Missouri, Nebraska, North Dakota, Ohio, South Dakota, Wisconsin)

Projected April 1 Fuel Price Indices (April 1, 2013 = 1.00)

Sector and Fuel	2029	2030	2031	2032	2033	2034	2035	2036	2037	2038	2039	2040	2041	2042	2043
Residential															
Electricity	1.08	1.08	1.08	1.08	1.09	1.09	1.10	1.10	1.11	1.12	1.12	1.13	1.13	1.13	1.13
Distillate Oil	1.13	1.15	1.17	1.18	1.20	1.22	1.25	1.27	1.30	1.32	1.34	1.36	1.38	1.41	1.43
LPG	1.12	1.13	1.14	1.14	1.15	1.16	1.17	1.18	1.20	1.21	1.22	1.23	1.24	1.25	1.27
Natural Gas	1.25	1.26	1.28	1.30	1.32	1.35	1.39	1.43	1.48	1.52	1.55	1.58	1.61	1.64	1.66
Commercial															
Electricity	1.06	1.06	1.06	1.06	1.07	1.08	1.09	1.10	1.11	1.13	1.14	1.14	1.15	1.15	1.16
Distillate Oil	1.06	1.08	1.10	1.11	1.13	1.15	1.18	1.20	1.22	1.24	1.27	1.29	1.31	1.34	1.36
Residual Oil	1.31	1.34	1.36	1.38	1.41	1.43	1.46	1.49	1.52	1.55	1.57	1.61	1.64	1.67	1.71
Natural Gas	1.27	1.29	1.30	1.32	1.34	1.37	1.41	1.46	1.51	1.56	1.59	1.63	1.66	1.69	1.73
Coal	1.12	1.12	1.13	1.14	1.14	1.15	1.17	1.17	1.18	1.19	1.20	1.21	1.22	1.23	1.24
Industrial															
Electricity	1.07	1.07	1.08	1.09	1.09	1.11	1.12	1.14	1.16	1.18	1.19	1.21	1.22	1.23	1.23
Distillate Oil	1.07	1.08	1.09	1.11	1.13	1.15	1.17	1.19	1.22	1.24	1.26	1.28	1.30	1.33	1.35
Residual Oil	1.36	1.39	1.41	1.44	1.46	1.49	1.52	1.55	1.58	1.60	1.63	1.67	1.70	1.74	1.77
Natural Gas	1.48	1.49	1.51	1.54	1.56	1.61	1.67	1.75	1.82	1.90	1.94	1.99	2.04	2.09	2.14
Coal	1.15	1.17	1.17	1.19	1.19	1.20	1.22	1.23	1.24	1.25	1.26	1.27	1.29	1.30	1.31
Transportation															
Motor Gasoline	1.09	1.10	1.11	1.13	1.14	1.16	1.18	1.20	1.22	1.25	1.27	1.30	1.32	1.34	1.37

32

Table Ca-3. Projected fuel price indices (excluding general inflation), by end-use sector and fuel type.

Census Region 3 (Alabama, Arkansas, Delaware, District of Columbia, Florida, Georgia, Kentucky, Louisiana, Maryland, Mississippi, North Carolina, Oklahoma, South Carolina, Tennessee, Texas, Virginia, West Virginia)

Projected April 1 Fuel Price Indices (April 1, 2013 = 1.00)

Sector and Fuel	2014	2015	2016	2017	2018	2019	2020	2021	2022	2023	2024	2025	2026	2027	2028
Residential															
Electricity	1.01	1.02	1.04	1.05	1.05	1.05	1.05	1.06	1.06	1.06	1.06	1.07	1.07	1.08	1.08
Distillate Oil	0.97	0.96	0.98	0.99	1.01	1.02	1.04	1.05	1.07	1.09	1.10	1.12	1.14	1.15	1.17
LPG	0.97	0.95	0.94	0.97	0.99	1.01	1.02	1.04	1.05	1.06	1.07	1.08	1.09	1.10	1.11
Natural Gas	0.98	0.96	0.99	1.01	1.04	1.06	1.07	1.08	1.10	1.13	1.14	1.16	1.17	1.18	1.20
Commercial															
Electricity	1.00	1.00	1.02	1.03	1.04	1.04	1.04	1.04	1.05	1.05	1.05	1.04	1.04	1.05	1.05
Distillate Oil	0.89	0.86	0.87	0.89	0.90	0.92	0.93	0.95	0.97	0.99	1.01	1.02	1.04	1.06	1.07
Residual Oil	0.85	0.80	0.81	0.82	0.84	0.86	0.88	0.89	0.91	0.93	0.95	0.97	0.99	1.01	1.03
Natural Gas	0.99	0.96	1.00	1.03	1.06	1.08	1.10	1.11	1.14	1.16	1.18	1.19	1.21	1.22	1.24
Coal	0.99	1.01	1.01	1.01	1.00	0.99	0.99	0.99	0.99	0.99	1.00	1.00	1.01	1.01	1.02
Industrial															
Electricity	1.00	1.00	1.02	1.04	1.05	1.05	1.05	1.06	1.07	1.07	1.08	1.08	1.09	1.10	1.11
Distillate Oil	0.90	0.88	0.89	0.91	0.93	0.94	0.96	0.98	1.00	1.02	1.03	1.05	1.06	1.08	1.09
Residual Oil	0.88	0.84	0.85	0.86	0.88	0.90	0.91	0.93	0.95	0.97	0.99	1.01	1.03	1.05	1.07
Natural Gas	1.06	1.08	1.18	1.25	1.31	1.34	1.36	1.39	1.44	1.50	1.54	1.56	1.60	1.62	1.65
Coal	0.99	1.01	1.02	1.03	1.03	1.03	1.03	1.04	1.04	1.05	1.05	1.06	1.06	1.07	1.07
Transportation															
Motor Gasoline	0.96	0.94	0.94	0.94	0.96	0.97	1.00	1.02	1.03	1.04	1.05	1.06	1.06	1.07	1.08

Table Ca-3, continued. Projected fuel price indices (excluding general inflation), by end-use sector and fuel type.

Census Region 3 (Alabama, Arkansas, Delaware, District of Columbia, Florida, Georgia, Kentucky, Louisiana, Maryland, Mississippi, North Carolina, Oklahoma, South Carolina, Tennessee, Texas, Virginia, West Virginia)

Projected April 1 Fuel Price Indices (April 1, 2013 = 1.00)

Sector and Fuel	2029	2030	2031	2032	2033	2034	2035	2036	2037	2038	2039	2040	2041	2042	2043
Residential															
Electricity	1.09	1.09	1.09	1.09	1.09	1.10	1.11	1.13	1.14	1.15	1.17	1.17	1.18	1.19	1.19
Distillate Oil	1.19	1.21	1.22	1.24	1.26	1.28	1.30	1.32	1.35	1.37	1.39	1.41	1.44	1.46	1.49
LPG	1.12	1.13	1.13	1.14	1.15	1.16	1.17	1.18	1.19	1.20	1.21	1.22	1.24	1.25	1.26
Natural Gas	1.21	1.23	1.24	1.26	1.27	1.30	1.33	1.37	1.40	1.44	1.46	1.48	1.50	1.53	1.55
Commercial															
Electricity	1.06	1.06	1.07	1.07	1.08	1.09	1.10	1.12	1.14	1.16	1.18	1.19	1.20	1.20	1.21
Distillate Oil	1.09	1.10	1.12	1.13	1.15	1.17	1.20	1.22	1.25	1.26	1.29	1.31	1.33	1.36	1.39
Residual Oil	1.05	1.08	1.09	1.11	1.13	1.15	1.18	1.21	1.24	1.25	1.28	1.31	1.34	1.38	1.41
Natural Gas	1.25	1.26	1.28	1.29	1.31	1.34	1.37	1.41	1.46	1.50	1.52	1.55	1.58	1.61	1.63
Coal	1.03	1.04	1.04	1.05	1.06	1.06	1.08	1.09	1.09	1.10	1.11	1.12	1.13	1.14	1.15
Industrial															
Electricity	1.12	1.12	1.13	1.14	1.15	1.17	1.19	1.22	1.24	1.27	1.29	1.31	1.32	1.34	1.35
Distillate Oil	1.11	1.12	1.14	1.15	1.17	1.19	1.21	1.24	1.26	1.28	1.30	1.33	1.35	1.37	1.40
Residual Oil	1.09	1.12	1.13	1.15	1.18	1.20	1.23	1.26	1.28	1.30	1.33	1.36	1.39	1.43	1.46
Natural Gas	1.68	1.70	1.73	1.75	1.79	1.86	1.93	2.04	2.15	2.25	2.30	2.36	2.43	2.50	2.58
Coal	1.08	1.09	1.10	1.11	1.11	1.12	1.13	1.14	1.15	1.16	1.17	1.18	1.19	1.20	1.21
Transportation															
Motor Gasoline	1.09	1.11	1.12	1.13	1.15	1.17	1.19	1.21	1.23	1.25	1.28	1.30	1.33	1.35	1.37

34

Table Ca-4. Projected fuel price indices (excluding general inflation), by end-use sector and fuel type.

Census Region 4 (Alaska, Arizona, California, Colorado, Hawaii,
Idaho, Montana, Nevada, New Mexico, Oregon, Utah, Washington, Wyoming)

Projected April 1 Fuel Price Indices (April 1, 2013 = 1.00)

Sector and Fuel	2014	2015	2016	2017	2018	2019	2020	2021	2022	2023	2024	2025	2026	2027	2028
Residential															
Electricity	1.00	1.00	1.01	1.01	1.01	1.00	1.00	1.00	0.99	0.99	0.99	0.99	0.99	0.99	0.99
Distillate Oil	0.91	0.89	0.91	0.92	0.94	0.95	0.97	0.99	1.00	1.02	1.03	1.05	1.07	1.08	1.10
LPG	0.97	0.94	0.94	0.96	0.99	1.00	1.02	1.04	1.05	1.06	1.07	1.08	1.09	1.10	1.11
Natural Gas	1.00	1.02	1.08	1.13	1.18	1.21	1.24	1.26	1.29	1.32	1.33	1.34	1.35	1.35	1.37
Commercial															
Electricity	1.00	1.00	1.01	1.02	1.01	1.01	1.01	1.02	1.01	1.00	0.99	1.00	1.00	1.00	0.99
Distillate Oil	0.88	0.85	0.86	0.88	0.90	0.92	0.93	0.95	0.97	0.98	1.00	1.02	1.03	1.05	1.07
Residual Oil	0.74	0.67	0.68	0.69	0.71	0.72	0.73	0.75	0.77	0.79	0.81	0.83	0.84	0.86	0.88
Natural Gas	1.01	1.01	1.06	1.11	1.17	1.21	1.23	1.26	1.29	1.32	1.34	1.34	1.35	1.36	1.37
Coal	1.01	1.03	1.05	1.05	1.06	1.07	1.08	1.09	1.10	1.11	1.12	1.13	1.14	1.17	1.18
Industrial															
Electricity	0.99	0.98	1.00	1.01	1.01	1.01	1.01	1.01	1.01	1.01	1.01	1.02	1.03	1.03	1.03
Distillate Oil	0.90	0.88	0.89	0.91	0.93	0.95	0.97	0.98	1.01	1.02	1.04	1.05	1.07	1.08	1.10
Residual Oil	0.77	0.71	0.71	0.73	0.75	0.77	0.78	0.80	0.82	0.84	0.88	0.90	0.92	0.94	0.96
Natural Gas	1.04	1.05	1.13	1.20	1.27	1.33	1.35	1.39	1.43	1.47	1.48	1.47	1.47	1.46	1.47
Coal	1.00	1.01	1.02	1.01	1.01	1.01	1.01	1.02	1.03	1.04	1.05	1.06	1.07	1.08	1.10
Transportation															
Motor Gasoline	0.97	0.95	0.95	0.96	0.98	1.00	1.02	1.04	1.05	1.06	1.07	1.08	1.09	1.09	1.10

Table Ca-4, continued. Projected fuel price indices (excluding general inflation), by end-use sector and fuel type.

Census Region 4 (Alaska, Arizona, California, Colorado, Hawaii,
Idaho, Montana, Nevada, New Mexico, Oregon, Utah, Washington, Wyoming)

Projected April 1 Fuel Price Indices (April 1, 2013 = 1.00)

Sector and Fuel	2029	2030	2031	2032	2033	2034	2035	2036	2037	2038	2039	2040	2041	2042	2043
Residential															
Electricity	1.00	1.00	1.01	1.02	1.03	1.03	1.04	1.06	1.07	1.08	1.09	1.11	1.12	1.12	1.13
Distillate Oil	1.12	1.14	1.15	1.17	1.19	1.21	1.23	1.25	1.27	1.29	1.31	1.33	1.36	1.38	1.40
LPG	1.11	1.12	1.13	1.13	1.14	1.15	1.16	1.17	1.18	1.19	1.20	1.21	1.23	1.24	1.25
Natural Gas	1.38	1.39	1.41	1.43	1.46	1.48	1.52	1.56	1.61	1.66	1.69	1.72	1.74	1.77	1.79
Commercial															
Electricity	0.99	0.99	1.00	1.01	1.01	1.01	1.03	1.04	1.05	1.07	1.08	1.10	1.11	1.11	1.11
Distillate Oil	1.08	1.10	1.11	1.13	1.15	1.17	1.19	1.21	1.24	1.26	1.28	1.30	1.32	1.35	1.37
Residual Oil	0.90	0.92	0.94	0.96	0.99	1.00	1.03	1.05	1.07	1.09	1.12	1.15	1.17	1.20	1.23
Natural Gas	1.38	1.39	1.41	1.43	1.46	1.49	1.53	1.57	1.62	1.68	1.71	1.74	1.77	1.80	1.83
Coal	1.20	1.22	1.24	1.26	1.28	1.30	1.32	1.34	1.36	1.38	1.40	1.42	1.43	1.44	1.45
Industrial															
Electricity	1.03	1.04	1.05	1.06	1.07	1.08	1.10	1.12	1.14	1.17	1.19	1.21	1.23	1.24	1.25
Distillate Oil	1.12	1.13	1.14	1.16	1.18	1.20	1.22	1.24	1.27	1.29	1.31	1.33	1.35	1.38	1.40
Residual Oil	0.98	1.00	1.03	1.05	1.07	1.10	1.13	1.14	1.18	1.21	1.24	1.26	1.30	1.33	1.37
Natural Gas	1.49	1.50	1.52	1.55	1.59	1.63	1.68	1.75	1.82	1.90	1.94	1.99	2.04	2.09	2.14
Coal	1.11	1.12	1.14	1.16	1.18	1.19	1.21	1.22	1.23	1.24	1.26	1.27	1.28	1.29	1.30
Transportation															
Motor Gasoline	1.12	1.13	1.14	1.15	1.17	1.19	1.21	1.23	1.25	1.27	1.30	1.32	1.34	1.37	1.39

Table Ca-5. Projected fuel price indices (excluding general inflation), by end-use sector and fuel type.

United States Average

Projected April 1 Fuel Price Indices (April 1, 2013 = 1.00)

Sector and Fuel	2014	2015	2016	2017	2018	2019	2020	2021	2022	2023	2024	2025	2026	2027	2028
Residential															
Electricity	1.00	1.01	1.02	1.03	1.03	1.03	1.03	1.03	1.03	1.03	1.03	1.04	1.04	1.05	1.05
Distillate Oil	0.97	0.97	0.98	1.00	1.01	1.03	1.04	1.06	1.07	1.09	1.11	1.12	1.14	1.16	1.18
LPG	0.97	0.94	0.94	0.96	0.99	1.01	1.02	1.04	1.05	1.06	1.08	1.09	1.10	1.10	1.11
Natural Gas	0.98	0.97	1.01	1.04	1.08	1.10	1.12	1.14	1.16	1.18	1.20	1.21	1.22	1.23	1.25
Commercial															
Electricity	1.00	0.99	1.01	1.02	1.02	1.02	1.02	1.03	1.03	1.03	1.02	1.02	1.02	1.02	1.02
Distillate Oil	0.90	0.88	0.89	0.91	0.93	0.94	0.96	0.97	0.99	1.01	1.03	1.05	1.06	1.08	1.09
Residual Oil	0.76	0.69	0.70	0.72	0.74	0.75	0.77	0.79	0.81	0.83	0.85	0.86	0.88	0.91	0.93
Natural Gas	0.99	0.97	1.01	1.05	1.09	1.11	1.13	1.14	1.17	1.19	1.21	1.22	1.23	1.24	1.25
Coal	0.99	1.01	1.02	1.02	1.04	1.04	1.05	1.06	1.06	1.07	1.08	1.08	1.10	1.10	1.11
Industrial															
Electricity	0.99	0.98	1.00	1.02	1.03	1.03	1.03	1.03	1.04	1.05	1.05	1.05	1.06	1.06	1.07
Distillate Oil	0.89	0.87	0.88	0.90	0.92	0.94	0.95	0.97	0.99	1.01	1.02	1.04	1.05	1.07	1.08
Residual Oil	0.87	0.83	0.84	0.85	0.87	0.89	0.90	0.92	0.94	0.96	0.98	1.00	1.02	1.04	1.06
Natural Gas	1.05	1.06	1.14	1.20	1.26	1.30	1.32	1.34	1.38	1.43	1.45	1.47	1.49	1.50	1.52
Coal	1.00	1.01	1.02	1.03	1.04	1.04	1.05	1.06	1.06	1.07	1.08	1.09	1.09	1.10	1.11
Transportation															
Motor Gasoline	0.96	0.95	0.94	0.95	0.97	0.98	1.01	1.02	1.04	1.05	1.06	1.06	1.07	1.08	1.09

Table Ca-5, continued. Projected fuel price indices (excluding general inflation), by end-use sector and fuel type.

United States Average

Projected April 1 Fuel Price Indices (April 1, 2013 = 1.00)

Sector and Fuel	2029	2030	2031	2032	2033	2034	2035	2036	2037	2038	2039	2040	2041	2042	2043
Residential															
Electricity	1.05	1.06	1.06	1.06	1.07	1.07	1.08	1.09	1.10	1.11	1.12	1.13	1.14	1.14	1.15
Distillate Oil	1.19	1.21	1.23	1.24	1.26	1.28	1.30	1.33	1.35	1.37	1.39	1.41	1.44	1.46	1.49
LPG	1.12	1.13	1.14	1.14	1.15	1.16	1.17	1.18	1.19	1.21	1.22	1.23	1.24	1.25	1.26
Natural Gas	1.26	1.27	1.29	1.31	1.33	1.35	1.39	1.42	1.46	1.50	1.53	1.55	1.58	1.60	1.63
Commercial															
Electricity	1.02	1.02	1.03	1.03	1.04	1.05	1.06	1.07	1.09	1.10	1.12	1.13	1.14	1.15	1.15
Distillate Oil	1.11	1.12	1.14	1.16	1.17	1.20	1.22	1.24	1.27	1.28	1.31	1.33	1.35	1.38	1.40
Residual Oil	0.95	0.97	0.98	1.01	1.03	1.05	1.08	1.11	1.14	1.16	1.19	1.22	1.25	1.28	1.31
Natural Gas	1.27	1.28	1.29	1.31	1.33	1.35	1.39	1.43	1.47	1.52	1.54	1.57	1.60	1.63	1.66
Coal	1.12	1.14	1.14	1.15	1.16	1.17	1.19	1.20	1.21	1.22	1.24	1.25	1.26	1.27	1.27
Industrial															
Electricity	1.07	1.08	1.09	1.10	1.11	1.12	1.14	1.16	1.18	1.21	1.23	1.25	1.26	1.27	1.28
Distillate Oil	1.10	1.11	1.13	1.14	1.16	1.18	1.20	1.23	1.25	1.27	1.29	1.31	1.34	1.36	1.39
Residual Oil	1.08	1.11	1.12	1.15	1.17	1.19	1.22	1.25	1.28	1.30	1.33	1.36	1.39	1.42	1.46
Natural Gas	1.55	1.56	1.59	1.61	1.65	1.70	1.77	1.85	1.93	2.01	2.06	2.11	2.17	2.23	2.29
Coal	1.12	1.13	1.14	1.15	1.16	1.17	1.18	1.20	1.22	1.22	1.23	1.24	1.25	1.26	1.27
Transportation															
Motor Gasoline	1.10	1.11	1.12	1.14	1.16	1.17	1.19	1.21	1.24	1.26	1.28	1.31	1.33	1.35	1.38

Table Cb-1. Projected average fuel price escalation rates, excluding general inflation, by end-use sector and fuel type.

Census Region 1 (Connecticut, Maine, Massachusetts, New Hampshire, New Jersey, New York, Pennsylvania, Rhode Island, Vermont)

Percentage change compounded annually

Sector and Fuel	2013 to 2018	2018 to 2023	2023 to 2028	2028 to 2033	2033 to 2038	2038 to 2043
Residential						
Electricity	-0.3	-0.3	1.4	0.5	0.6	0.8
Distillate Oil	0.4	1.5	1.5	1.4	1.6	1.7
LPG	-0.3	1.5	0.9	0.7	0.9	0.9
Natural Gas	0.9	1.4	1.1	1.1	2.0	1.4
Commercial						
Electricity	-0.5	0.1	-0.3	0.7	0.9	1.2
Distillate Oil	-0.7	1.8	1.5	1.4	1.8	1.8
Residual Oil	-6.2	2.4	2.4	2.1	2.3	2.6
Natural Gas	1.0	1.3	1.0	1.0	2.2	1.8
Coal	1.4	1.0	0.7	0.7	0.8	0.8
Industrial						
Electricity	-0.7	0.8	0.1	1.1	1.2	1.3
Distillate Oil	-1.1	1.8	1.5	1.3	1.8	1.8
Residual Oil	-5.1	2.3	2.4	2.1	2.3	2.6
Natural Gas	2.1	1.5	1.3	1.1	2.7	2.3
Coal	1.2	0.8	0.5	0.4	0.8	0.7
Transportation						
Motor Gasoline	-0.6	1.6	0.7	1.2	1.7	1.8

Table Cb-2. Projected average fuel price escalation rates, excluding general inflation, by end-use sector and fuel type.

Census Region 2 (Illinois, Indiana, Iowa, Kansas, Michigan, Minnesota, Missouri, Nebraska, North Dakota, Ohio, South Dakota, Wisconsin)

Percentage change compounded annually

Sector and Fuel	2013 to 2018	2018 to 2023	2023 to 2028	2028 to 2033	2033 to 2038	2038 to 2043
Residential						
Electricity	1.3	-0.0	0.3	0.1	0.6	0.3
Distillate Oil	-1.2	1.7	1.7	1.6	1.8	1.7
LPG	-0.2	1.5	0.9	0.7	0.9	1.0
Natural Gas	1.2	1.9	1.1	1.3	3.0	1.8
Commercial						
Electricity	1.1	0.1	-0.0	0.3	1.0	0.6
Distillate Oil	-2.6	1.9	1.7	1.5	1.9	1.8
Residual Oil	1.6	1.7	1.8	1.7	1.9	2.1
Natural Gas	1.7	2.0	1.0	1.2	3.2	2.0
Coal	0.6	0.8	0.6	0.6	0.9	0.8
Industrial						
Electricity	0.8	0.1	0.3	0.6	1.5	0.9
Distillate Oil	-2.3	1.9	1.5	1.4	1.9	1.8
Residual Oil	2.5	1.7	1.8	1.7	1.9	2.0
Natural Gas	4.1	2.4	1.2	1.4	3.9	2.4
Coal	0.8	0.9	0.9	0.9	1.0	0.9
Transportation						
Motor Gasoline	-0.8	1.6	0.7	1.2	1.7	1.9

Table Cb-3. Projected average fuel price escalation rates, excluding general inflation, by end-use sector and fuel type.

Census Region 3 (Alabama, Arkansas, Delaware, District of Columbia, Florida, Georgia, Kentucky, Louisiana, Maryland, Mississippi, North Carolina, Oklahoma, South Carolina, Tennessee, Texas, Virginia, West Virginia)

	Percentage change compounded annually					
Sector and Fuel	2013 to 2018	2018 to 2023	2023 to 2028	2028 to 2033	2033 to 2038	2038 to 2043
Residential						
Electricity	1.0	0.2	0.3	0.3	1.1	0.6
Distillate Oil	0.1	1.5	1.6	1.4	1.7	1.7
LPG	-0.2	1.5	0.9	0.7	0.9	0.9
Natural Gas	0.8	1.6	1.2	1.3	2.5	1.5
Commercial						
Electricity	0.8	0.2	0.0	0.4	1.5	0.9
Distillate Oil	-2.0	1.9	1.6	1.5	1.9	1.8
Residual Oil	-3.3	2.0	2.1	1.9	2.0	2.3
Natural Gas	1.2	1.8	1.2	1.1	2.8	1.7
Coal	0.1	-0.3	0.6	0.7	0.9	0.8
Industrial						
Electricity	0.9	0.5	0.6	0.8	2.0	1.2
Distillate Oil	-1.5	1.9	1.5	1.4	1.8	1.8
Residual Oil	-2.5	2.0	2.1	1.9	2.1	2.3
Natural Gas	5.5	2.8	1.9	1.7	4.7	2.8
Coal	0.6	0.3	0.5	0.7	0.9	0.7
Transportation						
Motor Gasoline	-0.9	1.7	0.8	1.2	1.7	1.9

Table Cb-4. Projected average fuel price escalation rates, excluding general inflation, by end-use sector and fuel type.

Census Region 4 (Alaska, Arizona, California, Colorado, Hawaii, Idaho, Montana, Nevada, New Mexico, Oregon, Utah, Washington, Wyoming)

Percentage change compounded annually

Sector and Fuel	2013 to 2018	2018 to 2023	2023 to 2028	2028 to 2033	2033 to 2038	2038 to 2043
Residential						
Electricity	0.2	-0.4	0.1	0.7	1.0	0.8
Distillate Oil	-1.2	1.6	1.6	1.5	1.7	1.7
LPG	-0.3	1.5	0.9	0.6	0.9	0.9
Natural Gas	3.3	2.3	0.8	1.3	2.6	1.6
Commercial						
Electricity	0.3	-0.3	-0.2	0.4	1.2	0.8
Distillate Oil	-2.1	1.8	1.6	1.5	1.8	1.8
Residual Oil	-6.7	2.1	2.3	2.3	2.1	2.4
Natural Gas	3.1	2.5	0.7	1.3	2.9	1.8
Coal	1.2	0.9	1.3	1.6	1.5	1.0
Industrial						
Electricity	0.3	-0.1	0.3	0.8	1.8	1.4
Distillate Oil	-1.4	1.9	1.5	1.3	1.8	1.7
Residual Oil	-5.6	2.4	2.5	2.2	2.5	2.4
Natural Gas	4.9	2.9	0.1	1.5	3.6	2.4
Coal	0.1	0.6	1.2	1.4	1.1	0.9
Transportation						
Motor Gasoline	-0.4	1.7	0.7	1.2	1.7	1.8

Table Cb-5. Projected average fuel price escalation rates, excluding general inflation, by end-use sector and fuel type.

United States Average

Percentage change compounded annually

Sector and Fuel	2013 to 2018	2018 to 2023	2023 to 2028	2028 to 2033	2033 to 2038	2038 to 2043
Residential						
Electricity	0.7	-0.1	0.4	0.3	0.9	0.6
Distillate Oil	0.2	1.5	1.5	1.4	1.6	1.7
LPG	-0.3	1.5	0.9	0.7	0.9	1.0
Natural Gas	1.5	1.8	1.1	1.2	2.5	1.6
Commercial						
Electricity	0.5	0.0	-0.1	0.4	1.2	0.9
Distillate Oil	-1.5	1.8	1.6	1.4	1.8	1.8
Residual Oil	-6.0	2.3	2.4	2.1	2.3	2.6
Natural Gas	1.7	1.9	1.0	1.1	2.7	1.8
Coal	0.7	0.7	0.8	0.9	1.0	0.8
Industrial						
Electricity	0.6	0.3	0.4	0.7	1.7	1.2
Distillate Oil	-1.7	1.9	1.5	1.4	1.8	1.8
Residual Oil	-2.8	2.0	2.1	1.9	2.1	2.3
Natural Gas	4.7	2.5	1.3	1.6	4.1	2.6
Coal	0.7	0.6	0.7	0.9	1.0	0.8
Transportation						
Motor Gasoline	-0.7	1.7	0.7	1.2	1.7	1.8

43

D. Projected Average Carbon Prices and Emissions Indices

Most financed federal projects, such as Energy Savings Performance Contracts (ESPC), base contract payments on projected annual energy cost savings. When setting up the contract, average rates of energy price escalation over the contract term are a matter of negotiation. One consideration in setting escalation rates is the potential for future carbon pricing. Should carbon pricing legislation be enacted by the U.S. Congress, use of the EIA-based escalation rates—which do not consider carbon pricing—likely would underestimate escalation for contract payments. To assist federal agencies in considering a range of escalation rate scenarios, in 2010 FEMP introduced to the Annual Supplement a new "D" series of tables projecting potential future carbon prices and electricity-related carbon emissions rates under a range of carbon policy scenarios. Average rates of escalation may be calculated for each of these carbon policy scenarios in the Energy Escalation Rate Calculator (EERC 2.0), a BLCC companion program for financed projects. These may be considered by federal agencies for use as energy price escalation rates for contract payments.

Carbon Price Projections. In January 2010, the U.S. Environmental Protection Agency (EPA) issued a supplemental analysis of the American Clean Energy and Security Act of 2009 (H.R. 2454).[3] The analysis was based on multiple peer-reviewed climate economic models, with a particular focus on an economy-wide model known as ADAGE (Applied Dynamic Analysis of Global Economy).

Three carbon policy scenarios are chosen from the EPA study to create three levels of carbon pricing: Default, Low, and High. Default Pricing stems from a scenario assuming H.R. 2454 is enacted as is with no changes in the policy design, which does not restrict the type of capacity that electric utilities may install to meet carbon emissions targets. The Default scenario assumes that all countries, including developing countries, will begin to restrict carbon emissions over the next 40 years. Low Pricing assumes that developing countries do not take any action over the next 40 years to restrict carbon emissions, which decreases the demand for carbon reductions and allows emitters in the United States to purchase carbon offsets from other countries at a lower cost. High Pricing assumes that carbon offsets are not allowed and nuclear and biomass capacity construction is restricted. Both assumptions limit some of the least expensive options available to decrease carbon emissions, causing carbon prices to increase.

Table D-1 presents projected U.S. carbon prices under the Low, Default, and High Pricing policy scenarios.

Carbon Emissions Projections. Electric utilities will adjust to carbon markets by changing their generation mixes. These adjustments—and resulting carbon emissions levels—will vary over time depending on the policy scenario, with higher carbon prices encouraging more rapid transformations to less carbon-intensive electricity generation. The supplemental EPA analysis reports ADAGE-projected emissions with no carbon policy (business as usual) and under the

---

[3]U.S. Environmental Protection Agency, *Supplemental EPA Analysis of the American Clean Energy and Security Act of 2009: H.R. 2454 in the 111th Congress*, Office of Atmospheric Programs, January 29, 2010, http://www.epa.gov/climatechange/economics/economicanalyses.html .

Low, Default, and High carbon pricing scenarios.[4] These emissions projections are used to develop carbon emissions rate indices for this report.

Table D-2 presents projected U.S. carbon emissions rate indices for electricity with no carbon policy and under each carbon policy scenario. These indices, when multiplied by annual electricity emissions as of the base date, provide estimates of future-year emissions.

Examples of How to Use the Carbon Prices and Electricity Emissions Indices:

For all fuel types except electricity: To compute the cost of carbon in 2040 assuming the default carbon pricing scenario, go to Table D-1, find the 2040 default carbon price ($0.07/kg), and multiply the price by total annual carbon dioxide emissions (in kg) as of the base date for all fuel types except electricity (project-specific emissions, by fuel type, are reported by the BLCC program). The result is expressed in constant 2013 dollars.

For electricity: To compute the cost of carbon in 2020 assuming high carbon pricing, make two calculations:
  (1) Go to Table D-1, find the 2020 High Pricing value ($0.09/kg), and multiply the price by annual carbon dioxide emissions for electricity (in kilograms) as of the base date.
  (2) Go to Table D-2, find the 2020 carbon dioxide emissions rate index for high carbon pricing (0.72), and multiply the index by the result in (1).
The result is the projected carbon cost for 2020 electricity consumption expressed in constant 2013 dollars.

---

[4] Emissions rates in the business-as-usual case change based on expected electricity capacity deployment and generator retirement given the current market and policy environment.

Table D-1. Projected Carbon Dioxide-Equivalent Emissions Prices, by Carbon Policy Scenario

(2013 dollars per kilogram carbon dioxide)

| Year | Carbon Policy Scenario | | |
	Default Pricing	Low Pricing	High Pricing
2013	$0.02	$0.01	$0.06
2014	$0.02	$0.01	$0.06
2015	$0.02	$0.01	$0.07
2016	$0.02	$0.01	$0.07
2017	$0.02	$0.02	$0.07
2018	$0.02	$0.02	$0.08
2019	$0.02	$0.02	$0.08
2020	$0.03	$0.02	$0.09
2021	$0.03	$0.02	$0.09
2022	$0.03	$0.02	$0.09
2023	$0.03	$0.02	$0.10
2024	$0.03	$0.02	$0.10
2025	$0.03	$0.02	$0.11
2026	$0.04	$0.02	$0.11
2027	$0.04	$0.03	$0.12
2028	$0.04	$0.03	$0.13
2029	$0.04	$0.03	$0.13
2030	$0.04	$0.03	$0.14
2031	$0.04	$0.03	$0.15
2032	$0.05	$0.03	$0.15
2033	$0.05	$0.03	$0.16
2034	$0.05	$0.04	$0.17
2035	$0.05	$0.04	$0.18
2036	$0.06	$0.04	$0.19
2037	$0.06	$0.04	$0.19
2038	$0.06	$0.04	$0.20
2039	$0.07	$0.05	$0.21
2040	$0.07	$0.05	$0.23
2041	$0.07	$0.05	$0.24
2042	$0.08	$0.05	$0.25
2043	$0.08	$0.06	$0.26

Table D-2. Projected Carbon Dioxide Emissions Rate Indices for Electricity, by Carbon Policy Scenario

Year	Carbon Policy Scenario			
	No Policy	Default Pricing	Low Pricing	High Pricing
2013	1.00	0.94	0.94	0.92
2014	0.99	0.92	0.92	0.90
2015	0.99	0.90	0.90	0.87
2016	0.98	0.88	0.88	0.84
2017	0.98	0.85	0.85	0.81
2018	0.97	0.82	0.82	0.78
2019	0.96	0.80	0.79	0.75
2020	0.96	0.77	0.77	0.72
2021	0.95	0.75	0.75	0.70
2022	0.95	0.73	0.73	0.67
2023	0.95	0.71	0.71	0.65
2024	0.94	0.69	0.69	0.63
2025	0.94	0.67	0.67	0.61
2026	0.94	0.64	0.64	0.59
2027	0.94	0.62	0.62	0.57
2028	0.95	0.60	0.59	0.55
2029	0.95	0.57	0.57	0.53
2030	0.95	0.55	0.54	0.52
2031	0.95	0.53	0.53	0.49
2032	0.95	0.51	0.51	0.47
2033	0.95	0.50	0.49	0.45
2034	0.95	0.48	0.47	0.43
2035	0.95	0.46	0.45	0.41
2036	0.95	0.44	0.43	0.39
2037	0.95	0.42	0.41	0.36
2038	0.95	0.40	0.39	0.34
2039	0.95	0.38	0.37	0.31
2040	0.95	0.36	0.35	0.29
2041	0.95	0.34	0.33	0.26
2042	0.95	0.32	0.31	0.22
2043	0.96	0.29	0.28	0.19

PART II:
ENERGY PRICE INDICES FOR PRIVATE SECTOR LCC ANALYSIS

This section presents tables of projected nominal (i.e., including inflation) fuel price indices for four fuels in the residential sector and five fuels in the commercial sector for each of the years from 2013 through 2043. These price indices are based on the DOE energy price projections, reported in Part I, Section B of this document, used to calculate the FEMP and OMB UPV* factors for energy costs. Tables S-1 to S-5 are provided as an update to similar tables originally published in *Comprehensive Guide for Least-Cost Energy Decisions* (NBS SP 709).

As a convenience for the user, the indices include the effect of four alternative, hypothetical rates of general price inflation: 2 %, 3 %, 4 %, and 5 %. Selection of these rates is in no way intended to suggest what actual rates might be. Use of the indices produce price estimates in current dollars, inclusive of general price inflation. Current-dollar prices are needed when discounting is performed with discount rates that include general price inflation (i.e., *nominal* or *market* discount rates).

The calculated indices with inflation rates of 2 %, 3 %, 4 %, and 5 % allow the analyst to perform evaluations based on the assumption of a positive rate of general price inflation that changes the purchasing power of the dollar. Performing evaluations in current dollars is sometimes preferred for private investment decisions, primarily because it facilitates the treatment of taxes.

The indices in Tables S-1 through S-5 are derived from the indices reported in Tables Ca-1 through Ca-5 by means of the following equation:

$$I_S = I_C \times (1 + g)^N,$$

where I_S = index found in Tables S-1 through S-5;
$\quad\quad I_C$ = index found in Tables Ca-1 through Ca-5;
$\quad\quad g$ = annual rate of general price inflation in decimal form; and
$\quad\quad N$ = number of years, in this case equal to the year of the index minus 2013.

Example of How to Use the Indices:

Suppose you wish to estimate the annual cost of natural gas for a house in Maryland in year 2016, given the annual cost in 2013 (April 1) prices, and you expect an annual inflation rate of 3 % per year. From table S-3, find the column with residential natural gas indices at an inflation rate of 3 %; then locate the index for the year 2016. This index is 1.08. Multiply the annual cost in 2013 prices by the index to find the estimated annual cost in year 2016 prices.

If this annual cost in year-2016 prices is to be discounted to present value, you must use a nominal discount rate that includes the same assumption with regard to general price inflation (3 % in this example). To obtain a present-value cost over the entire study period, the present-value calculation must be repeated for each year that there are natural gas costs, and the results summed. (UPV* factors are not given for private sector use because of the large number of tables required to cover potential discount rates that might be used by the analyst.) The BLCC computer program can perform LCC analyses using any discount rate, in constant or in current (market) dollars. The private sector analyst may use the UPV* factors reported in Part I, provided the analysis is performed in constant dollars and the desired discount rate corresponds to the DOE or OMB discount rates used in Part I.

For further explanation of the use of these indices, see NBS Special Publication 709, appendix B, Part I.

The data in the tables that follow are reported for the four Census regions and the U.S. average. Figure B-1 on page 13 presents a map showing the states corresponding to the four Census regions. The Census regions do not include American Samoa, Canal Zone, Guam, Puerto Rico, Trust Territory of the Pacific Islands, or the Virgin Islands. Analysts of federal projects in these areas should use data which are "reasonable under the circumstances," and may refer to the tables with U.S. average data for guidance.

Table S-1. Projected fuel price indices with assumed general price inflation rates of 2 %, 3 %, 4 %, and 5 %, by end-use sector and fuel type.

Census Region 1 (Connecticut, Maine, Massachusetts, New Hampshire, New Jersey, New York, Pennsylvania, Rhode Island, Vermont)

Projected April 1 Fuel Price Indices (April 1, 2013 = 1.00)

-----------------------------RESIDENTIAL-----------------------------

Year	Electricity				Distillate Oil				LPG				Natural Gas			
	Inflation Rate				Inflation Rate				Inflation Rate				Inflation Rate			
	2%	3%	4%	5%	2%	3%	4%	5%	2%	3%	4%	5%	2%	3%	4%	5%
2014	1.01	1.02	1.03	1.04	1.00	1.01	1.02	1.03	0.99	1.00	1.01	1.02	1.01	1.02	1.03	1.04
2015	1.03	1.05	1.07	1.09	1.02	1.04	1.06	1.08	0.98	1.00	1.02	1.04	1.01	1.03	1.05	1.07
2016	1.05	1.08	1.12	1.15	1.05	1.08	1.11	1.14	1.00	1.03	1.06	1.09	1.06	1.09	1.12	1.15
2017	1.07	1.11	1.16	1.20	1.09	1.13	1.18	1.22	1.04	1.09	1.13	1.17	1.11	1.15	1.19	1.24
2018	1.09	1.14	1.20	1.26	1.12	1.18	1.24	1.30	1.09	1.14	1.20	1.26	1.16	1.21	1.27	1.34
2019	1.10	1.17	1.24	1.31	1.16	1.23	1.31	1.39	1.13	1.20	1.27	1.35	1.20	1.27	1.35	1.43
2020	1.12	1.20	1.28	1.37	1.20	1.29	1.38	1.48	1.18	1.26	1.35	1.44	1.24	1.33	1.42	1.52
2021	1.14	1.23	1.33	1.43	1.25	1.35	1.46	1.57	1.22	1.31	1.42	1.53	1.28	1.38	1.49	1.61
2022	1.15	1.26	1.37	1.50	1.29	1.41	1.54	1.68	1.26	1.37	1.50	1.63	1.32	1.44	1.57	1.71
2023	1.18	1.31	1.44	1.58	1.34	1.47	1.62	1.79	1.30	1.43	1.57	1.73	1.37	1.51	1.66	1.83
2024	1.22	1.36	1.51	1.68	1.38	1.54	1.71	1.90	1.34	1.49	1.65	1.84	1.41	1.57	1.75	1.94
2025	1.26	1.42	1.59	1.79	1.44	1.61	1.81	2.03	1.38	1.55	1.74	1.95	1.45	1.63	1.83	2.06
2026	1.30	1.48	1.67	1.90	1.49	1.69	1.91	2.17	1.42	1.61	1.82	2.06	1.51	1.71	1.94	2.20
2027	1.35	1.55	1.77	2.03	1.54	1.76	2.02	2.31	1.46	1.67	1.91	2.18	1.55	1.78	2.03	2.33
2028	1.40	1.62	1.87	2.16	1.59	1.84	2.13	2.46	1.50	1.73	2.00	2.31	1.60	1.85	2.14	2.47
2029	1.43	1.67	1.95	2.28	1.65	1.93	2.25	2.62	1.54	1.80	2.10	2.44	1.64	1.92	2.24	2.61
2030	1.46	1.73	2.04	2.40	1.70	2.01	2.37	2.79	1.58	1.86	2.20	2.58	1.69	2.00	2.35	2.77
2031	1.50	1.79	2.13	2.53	1.76	2.10	2.50	2.97	1.62	1.93	2.30	2.73	1.75	2.08	2.48	2.94
2032	1.54	1.85	2.23	2.67	1.82	2.19	2.63	3.16	1.66	2.00	2.40	2.88	1.80	2.17	2.61	3.13
2033	1.59	1.93	2.34	2.83	1.88	2.29	2.78	3.36	1.71	2.08	2.52	3.05	1.86	2.26	2.74	3.32
2034	1.61	1.98	2.43	2.97	1.95	2.39	2.93	3.58	1.76	2.15	2.64	3.23	1.93	2.37	2.90	3.54
2035	1.66	2.05	2.54	3.13	2.02	2.51	3.10	3.83	1.81	2.24	2.77	3.42	2.00	2.48	3.07	3.79
2036	1.72	2.15	2.68	3.34	2.10	2.63	3.28	4.09	1.86	2.33	2.91	3.62	2.09	2.61	3.26	4.07
2037	1.77	2.23	2.81	3.54	2.18	2.75	3.47	4.37	1.92	2.42	3.06	3.84	2.17	2.74	3.46	4.35
2038	1.81	2.31	2.94	3.73	2.25	2.88	3.66	4.65	1.98	2.52	3.21	4.08	2.26	2.89	3.68	4.67
2039	1.88	2.42	3.11	3.99	2.34	3.01	3.87	4.96	2.03	2.62	3.37	4.32	2.34	3.02	3.88	4.97
2040	1.93	2.51	3.26	4.22	2.42	3.15	4.09	5.30	2.09	2.72	3.53	4.57	2.43	3.16	4.10	5.31
2041	1.98	2.60	3.40	4.45	2.51	3.30	4.32	5.65	2.15	2.83	3.71	4.85	2.51	3.30	4.32	5.65
2042	2.03	2.69	3.56	4.70	2.60	3.46	4.57	6.04	2.22	2.94	3.90	5.14	2.60	3.44	4.56	6.02
2043	2.08	2.79	3.72	4.96	2.70	3.62	4.84	6.45	2.29	3.06	4.09	5.46	2.68	3.60	4.81	6.40

Table S-1, continued. Projected fuel price indices with assumed general price inflation rates of 2 %, 3 %, 4 %, and 5 %, by end-use sector and fuel type.

Census Region 1 (Connecticut, Maine, Massachusetts, New Hampshire, New Jersey, New York, Pennsylvania, Rhode Island, Vermont)

Projected April 1 Fuel Price Indices (April 1, 2013 = 1.00)

--COMMERCIAL--

Year	Electricity Inflation Rate 2 %	3 %	4 %	5 %	Distillate Oil Inflation Rate 2 %	3 %	4 %	5 %	Residual Oil Inflation Rate 2 %	3 %	4 %	5 %	Natural Gas Inflation Rate 2 %	3 %	4 %	5 %	Coal Inflation Rate 2 %	3 %	4 %	5 %
2014	1.00	1.01	1.02	1.03	0.95	0.96	0.97	0.98	0.76	0.77	0.78	0.79	1.01	1.02	1.03	1.04	1.02	1.03	1.04	1.05
2015	1.00	1.02	1.04	1.06	0.96	0.98	0.99	1.01	0.71	0.72	0.74	0.75	1.01	1.03	1.05	1.07	1.05	1.07	1.09	1.11
2016	1.02	1.05	1.08	1.11	0.99	1.02	1.05	1.08	0.74	0.76	0.78	0.80	1.06	1.09	1.12	1.16	1.08	1.11	1.15	1.18
2017	1.05	1.09	1.13	1.18	1.03	1.07	1.11	1.15	0.76	0.79	0.83	0.86	1.11	1.15	1.20	1.25	1.12	1.16	1.21	1.26
2018	1.08	1.13	1.19	1.25	1.07	1.11	1.18	1.23	0.80	0.84	0.88	0.93	1.16	1.22	1.28	1.34	1.18	1.24	1.30	1.37
2019	1.11	1.18	1.25	1.32	1.11	1.17	1.24	1.32	0.83	0.88	0.94	0.99	1.21	1.28	1.35	1.43	1.23	1.30	1.38	1.46
2020	1.13	1.21	1.30	1.38	1.15	1.23	1.31	1.40	0.87	0.93	1.00	1.07	1.24	1.33	1.42	1.52	1.27	1.36	1.45	1.55
2021	1.15	1.24	1.34	1.45	1.19	1.29	1.39	1.50	0.91	0.98	1.06	1.15	1.28	1.38	1.49	1.61	1.30	1.41	1.52	1.64
2022	1.17	1.27	1.39	1.52	1.24	1.35	1.48	1.61	0.95	1.04	1.13	1.24	1.32	1.44	1.57	1.72	1.34	1.46	1.60	1.74
2023	1.20	1.32	1.46	1.60	1.28	1.42	1.56	1.72	0.99	1.10	1.21	1.33	1.37	1.51	1.66	1.83	1.38	1.52	1.67	1.84
2024	1.23	1.37	1.52	1.69	1.33	1.48	1.65	1.83	1.04	1.16	1.29	1.43	1.41	1.57	1.75	1.94	1.41	1.57	1.75	1.95
2025	1.23	1.39	1.56	1.75	1.38	1.55	1.74	1.95	1.08	1.22	1.37	1.53	1.45	1.63	1.83	2.05	1.45	1.63	1.83	2.05
2026	1.25	1.44	1.58	1.79	1.43	1.62	1.84	2.08	1.13	1.28	1.45	1.65	1.50	1.71	1.94	2.19	1.49	1.69	1.92	2.17
2027	1.30	1.51	1.64	1.88	1.48	1.70	1.94	2.22	1.18	1.36	1.55	1.77	1.54	1.77	2.03	2.32	1.53	1.75	2.01	2.30
2028	1.34	1.56	1.74	2.01	1.53	1.77	2.05	2.36	1.24	1.43	1.66	1.91	1.59	1.84	2.12	2.45	1.57	1.82	2.10	2.43
2029	1.37	1.62	1.82	2.13	1.58	1.85	2.16	2.52	1.29	1.51	1.76	2.05	1.63	1.91	2.23	2.60	1.62	1.89	2.21	2.57
2030	1.40	1.67	1.91	2.24	1.64	1.93	2.28	2.68	1.35	1.59	1.88	2.21	1.68	1.98	2.33	2.75	1.66	1.96	2.31	2.72
2031	1.44	1.73	1.99	2.37	1.69	2.02	2.40	2.85	1.39	1.66	1.97	2.34	1.73	2.06	2.46	2.92	1.71	2.04	2.42	2.88
2032	1.49	1.81	2.08	2.49	1.75	2.10	2.53	3.03	1.45	1.75	2.10	2.52	1.78	2.14	2.58	3.09	1.75	2.11	2.53	3.04
2033	1.52	1.86	2.19	2.66	1.81	2.20	2.67	3.23	1.52	1.85	2.24	2.71	1.84	2.23	2.71	3.28	1.80	2.18	2.65	3.21
2034	1.56	1.93	2.28	2.79	1.88	2.31	2.82	3.45	1.58	1.94	2.37	2.90	1.91	2.34	2.86	3.50	1.84	2.26	2.77	3.39
2035	1.61	2.02	2.39	2.95	1.95	2.42	2.99	3.69	1.66	2.05	2.54	3.14	1.98	2.45	3.03	3.75	1.90	2.35	2.91	3.59
2036	1.66	2.10	2.52	3.14	2.03	2.54	3.17	3.95	1.73	2.17	2.71	3.38	2.07	2.59	3.24	4.04	1.96	2.45	3.06	3.81
2037	1.72	2.19	2.64	3.33	2.11	2.66	3.36	4.23	1.81	2.29	2.89	3.64	2.16	2.73	3.44	4.33	2.01	2.54	3.20	4.03
2038	1.81	2.33	2.79	3.54	2.18	2.78	3.55	4.50	1.88	2.40	3.06	3.88	2.26	2.88	3.67	4.66	2.07	2.64	3.36	4.26
2039	1.87	2.43	2.99	3.84	2.26	2.92	3.75	4.81	1.97	2.54	3.27	4.19	2.34	3.01	3.87	4.96	2.13	2.74	3.53	4.52
2040	1.91	2.52	3.15	4.08	2.35	3.06	3.97	5.14	2.06	2.68	3.48	4.51	2.42	3.16	4.10	5.30	2.19	2.85	3.69	4.78
2041	1.96	2.60	3.30	4.31	2.44	3.20	4.20	5.49	2.16	2.83	3.71	4.85	2.52	3.31	4.34	5.67	2.25	2.95	3.87	5.06
2042	2.01	2.70	3.45	4.55	2.53	3.36	4.45	5.87	2.26	2.99	3.96	5.23	2.62	3.47	4.60	6.07	2.31	3.07	4.06	5.36
2043	2.06	2.80	3.60	4.80	2.63	3.52	4.71	6.27	2.36	3.17	4.23	5.64	2.72	3.64	4.87	6.49	2.38	3.19	4.26	5.67

Table S-1, continued. Projected fuel price indices with assumed general price inflation rates of 2 %, 3 %, 4 %, and 5 %, by end-use sector and fuel type.

Census Region 1 (Connecticut, Maine, Massachusetts, New Hampshire,
New Jersey, New York, Pennsylvania, Rhode Island, Vermont)

Projected April 1 Fuel Price Indices (April 1, 2013 = 1.00)

-----INDUSTRIAL-----

Year	Electricity 2 %	3 %	4 %	5 %	Distillate Oil 2 %	3 %	4 %	5 %	Residual Oil 2 %	3 %	4 %	5 %	Natural Gas 2 %	3 %	4 %	5 %	Coal 2 %	3 %	4 %	5 %
2014	1.00	1.01	1.01	1.02	0.94	0.94	0.95	0.96	0.80	0.81	0.81	0.82	1.02	1.03	1.04	1.05	1.01	1.02	1.03	1.04
2015	0.99	1.01	1.03	1.05	0.93	0.95	0.97	0.99	0.75	0.77	0.78	0.80	1.04	1.06	1.08	1.10	1.05	1.08	1.10	1.12
2016	1.02	1.05	1.08	1.11	0.97	0.99	1.02	1.05	0.78	0.81	0.83	0.85	1.10	1.14	1.17	1.20	1.10	1.13	1.16	1.19
2017	1.04	1.08	1.12	1.17	1.01	1.05	1.09	1.13	0.81	0.84	0.88	0.91	1.16	1.21	1.25	1.30	1.13	1.17	1.22	1.26
2018	1.07	1.12	1.18	1.23	1.04	1.10	1.15	1.21	0.85	0.89	0.94	0.98	1.22	1.29	1.35	1.41	1.17	1.23	1.29	1.35
2019	1.09	1.16	1.23	1.30	1.09	1.15	1.22	1.29	0.88	0.94	0.99	1.05	1.27	1.35	1.43	1.52	1.20	1.28	1.35	1.43
2020	1.12	1.20	1.28	1.37	1.12	1.20	1.29	1.38	0.92	0.99	1.06	1.13	1.32	1.41	1.51	1.61	1.24	1.33	1.43	1.52
2021	1.15	1.24	1.34	1.45	1.17	1.26	1.36	1.47	0.96	1.04	1.13	1.21	1.35	1.46	1.58	1.70	1.27	1.38	1.49	1.60
2022	1.18	1.29	1.40	1.53	1.22	1.33	1.45	1.58	1.01	1.10	1.20	1.30	1.40	1.53	1.67	1.82	1.31	1.43	1.56	1.70
2023	1.23	1.35	1.49	1.64	1.26	1.39	1.53	1.69	1.05	1.16	1.27	1.40	1.46	1.61	1.77	1.95	1.35	1.48	1.63	1.80
2024	1.25	1.40	1.55	1.72	1.31	1.46	1.62	1.80	1.10	1.22	1.36	1.51	1.51	1.68	1.87	2.07	1.38	1.54	1.71	1.90
2025	1.27	1.42	1.60	1.79	1.36	1.52	1.71	1.92	1.14	1.28	1.44	1.62	1.55	1.74	1.96	2.20	1.41	1.59	1.79	2.00
2026	1.28	1.45	1.64	1.86	1.40	1.59	1.80	2.04	1.19	1.35	1.53	1.74	1.62	1.84	2.08	2.36	1.45	1.65	1.87	2.12
2027	1.32	1.51	1.73	1.98	1.45	1.66	1.90	2.18	1.25	1.43	1.63	1.87	1.67	1.91	2.19	2.50	1.49	1.71	1.95	2.23
2028	1.36	1.58	1.82	2.10	1.50	1.74	2.01	2.32	1.30	1.51	1.74	2.01	1.71	1.98	2.29	2.65	1.53	1.77	2.04	2.36
2029	1.40	1.63	1.91	2.22	1.55	1.81	2.11	2.46	1.36	1.59	1.85	2.16	1.77	2.07	2.41	2.81	1.56	1.82	2.13	2.48
2030	1.44	1.70	2.00	2.36	1.60	1.89	2.23	2.62	1.42	1.68	1.97	2.32	1.82	2.15	2.53	2.98	1.60	1.89	2.22	2.62
2031	1.48	1.77	2.10	2.50	1.65	1.97	2.34	2.78	1.46	1.74	2.07	2.46	1.88	2.24	2.67	3.17	1.64	1.95	2.32	2.76
2032	1.53	1.84	2.21	2.65	1.71	2.05	2.47	2.96	1.53	1.84	2.21	2.65	1.94	2.33	2.80	3.36	1.68	2.02	2.43	2.91
2033	1.58	1.93	2.34	2.83	1.77	2.15	2.61	3.16	1.60	1.94	2.35	2.85	2.00	2.44	2.95	3.58	1.72	2.09	2.53	3.06
2034	1.60	1.96	2.40	2.93	1.84	2.25	2.76	3.37	1.65	2.03	2.49	3.04	2.09	2.56	3.14	3.83	1.76	2.16	2.64	3.23
2035	1.64	2.03	2.52	3.11	1.91	2.36	2.92	3.61	1.74	2.15	2.66	3.29	2.18	2.70	3.34	4.12	1.81	2.25	2.78	3.43
2036	1.73	2.17	2.71	3.37	1.98	2.48	3.10	3.86	1.82	2.27	2.84	3.54	2.29	2.87	3.58	4.47	1.87	2.34	2.92	3.64
2037	1.80	2.27	2.87	3.61	2.06	2.60	3.28	4.13	1.90	2.40	3.03	3.81	2.40	3.03	3.82	4.80	1.92	2.42	3.06	3.85
2038	1.86	2.37	3.02	3.84	2.13	2.72	3.47	4.40	1.97	2.51	3.20	4.07	2.52	3.22	4.10	5.21	1.97	2.51	3.20	4.07
2039	1.97	2.53	3.26	4.18	2.21	2.85	3.67	4.70	2.06	2.66	3.42	4.38	2.62	3.38	4.35	5.57	2.03	2.61	3.36	4.31
2040	2.07	2.63	3.41	4.42	2.30	2.99	3.88	5.02	2.15	2.80	3.64	4.71	2.73	3.55	4.61	5.97	2.08	2.71	3.52	4.55
2041	2.07	2.72	3.57	4.67	2.38	3.13	4.11	5.37	2.25	2.96	3.88	5.07	2.85	3.75	4.91	6.42	2.14	2.81	3.68	4.81
2042	2.13	2.83	3.75	4.95	2.47	3.28	4.35	5.74	2.36	3.13	4.14	5.47	2.98	3.96	5.24	6.91	2.20	2.92	3.86	5.10
2043	2.20	2.94	3.93	5.24	2.57	3.44	4.60	6.13	2.47	3.31	4.42	5.89	3.11	4.17	5.58	7.43	2.26	3.02	4.04	5.39

Table S-2. Projected fuel price indices with assumed general price inflation rates of 2 %, 3 %, 4 %, and 5 %, by end-use sector and fuel type.

Census Region 2 (Illinois, Indiana, Iowa, Kansas, Michigan, Minnesota, Missouri, Nebraska, North Dakota, Ohio, South Dakota, Wisconsin)

Projected April 1 Fuel Price Indices (April 1, 2013 = 1.00)

----RESIDENTIAL----

Year	Electricity Inflation Rate				Distillate Oil Inflation Rate				LPG Inflation Rate				Natural Gas Inflation Rate			
	2 %	3 %	4 %	5 %	2 %	3 %	4 %	5 %	2 %	3 %	4 %	5 %	2 %	3 %	4 %	5 %
2014	1.02	1.03	1.04	1.05	0.94	0.95	0.96	0.97	0.99	1.00	1.01	1.02	0.99	1.00	1.01	1.02
2015	1.06	1.08	1.10	1.12	0.94	0.96	0.97	0.99	0.98	1.00	1.02	1.04	0.98	1.00	1.02	1.04
2016	1.10	1.13	1.17	1.20	0.97	1.00	1.03	1.06	1.00	1.03	1.06	1.09	1.04	1.07	1.10	1.13
2017	1.15	1.19	1.24	1.29	1.01	1.05	1.09	1.13	1.05	1.09	1.13	1.17	1.10	1.15	1.19	1.24
2018	1.18	1.24	1.30	1.36	1.04	1.09	1.15	1.20	1.09	1.15	1.20	1.26	1.17	1.23	1.29	1.36
2019	1.20	1.27	1.35	1.43	1.08	1.15	1.21	1.29	1.13	1.20	1.27	1.35	1.23	1.30	1.38	1.46
2020	1.22	1.30	1.40	1.49	1.12	1.20	1.28	1.37	1.18	1.26	1.35	1.44	1.27	1.36	1.46	1.56
2021	1.24	1.34	1.45	1.57	1.16	1.26	1.36	1.46	1.22	1.32	1.42	1.54	1.32	1.43	1.54	1.66
2022	1.27	1.39	1.51	1.65	1.20	1.31	1.43	1.56	1.26	1.37	1.50	1.63	1.37	1.50	1.63	1.78
2023	1.30	1.43	1.58	1.74	1.25	1.38	1.52	1.67	1.30	1.43	1.58	1.74	1.43	1.57	1.73	1.91
2024	1.33	1.49	1.65	1.84	1.29	1.44	1.60	1.78	1.34	1.49	1.66	1.84	1.48	1.64	1.83	2.03
2025	1.37	1.54	1.73	1.93	1.34	1.51	1.70	1.90	1.38	1.55	1.74	1.95	1.52	1.71	1.92	2.15
2026	1.40	1.59	1.80	2.04	1.39	1.58	1.80	2.03	1.42	1.61	1.83	2.07	1.57	1.78	2.02	2.28
2027	1.43	1.64	1.87	2.14	1.45	1.66	1.90	2.17	1.46	1.67	1.92	2.19	1.61	1.85	2.12	2.42
2028	1.46	1.69	1.95	2.25	1.50	1.73	2.00	2.31	1.50	1.74	2.01	2.32	1.66	1.92	2.22	2.57
2029	1.49	1.74	2.03	2.36	1.55	1.81	2.12	2.47	1.54	1.80	2.10	2.45	1.72	2.01	2.34	2.73
2030	1.52	1.79	2.11	2.48	1.61	1.90	2.24	2.64	1.58	1.87	2.20	2.59	1.77	2.09	2.46	2.90
2031	1.55	1.85	2.20	2.61	1.67	1.99	2.36	2.81	1.62	1.94	2.30	2.74	1.83	2.18	2.59	3.08
2032	1.58	1.90	2.29	2.74	1.72	2.08	2.49	2.99	1.67	2.01	2.41	2.89	1.89	2.27	2.73	3.27
2033	1.61	1.96	2.38	2.88	1.79	2.17	2.63	3.19	1.71	2.08	2.53	3.06	1.96	2.38	2.89	3.50
2034	1.65	2.03	2.48	3.04	1.86	2.28	2.79	3.41	1.76	2.16	2.65	3.24	2.05	2.51	3.07	3.76
2035	1.69	2.10	2.60	3.20	1.93	2.39	2.96	3.65	1.81	2.25	2.78	3.43	2.14	2.66	3.29	4.06
2036	1.74	2.18	2.72	3.39	2.01	2.51	3.14	3.91	1.87	2.34	2.92	3.64	2.26	2.83	3.53	4.40
2037	1.79	2.26	2.85	3.59	2.08	2.63	3.32	4.18	1.92	2.43	3.07	3.86	2.38	3.01	3.79	4.77
2038	1.83	2.34	2.98	3.79	2.16	2.76	3.51	4.46	1.98	2.53	3.22	4.09	2.50	3.19	4.06	5.16
2039	1.88	2.42	3.11	3.99	2.24	2.88	3.71	4.76	2.04	2.63	3.38	4.33	2.60	3.35	4.31	5.52
2040	1.92	2.50	3.25	4.20	2.32	3.02	3.92	5.08	2.10	2.73	3.55	4.59	2.70	3.51	4.56	5.91
2041	1.96	2.58	3.38	4.42	2.41	3.16	4.15	5.42	2.16	2.84	3.72	4.87	2.80	3.68	4.83	6.31
2042	2.01	2.67	3.53	4.66	2.50	3.32	4.39	5.80	2.23	2.96	3.91	5.16	2.91	3.86	5.10	6.74
2043	2.05	2.75	3.68	4.90	2.60	3.48	4.65	6.20	2.30	3.08	4.11	5.48	3.01	4.04	5.40	7.19

Table S-2, continued. Projected fuel price indices with assumed general price inflation rates of 2 %, 3 %, 4 %, and 5 %, by end-use sector and fuel type.

Census Region 2 (Illinois, Indiana, Iowa, Kansas, Michigan, Minnesota, Missouri, Nebraska, North Dakota, Ohio, South Dakota, Wisconsin)

Projected April 1 Fuel Price Indices (April 1, 2013 = 1.00)

----------COMMERCIAL----------

Year	Electricity				Distillate Oil				Residual Oil				Natural Gas				Coal			
	Inflation Rate				Inflation Rate				Inflation Rate				Inflation Rate				Inflation Rate			
	2 %	3 %	4 %	5 %	2 %	3 %	4 %	5 %	2 %	3 %	4 %	5 %	2 %	3 %	4 %	5 %	2 %	3 %	4 %	5 %
2014	1.01	1.02	1.03	1.04	0.88	0.89	0.90	0.91	1.06	1.07	1.08	1.10	1.01	1.02	1.02	1.03	1.01	1.02	1.03	1.04
2015	1.04	1.06	1.08	1.10	0.87	0.88	0.90	0.92	1.09	1.11	1.13	1.15	1.00	1.01	1.03	1.05	1.04	1.06	1.08	1.10
2016	1.08	1.11	1.14	1.18	0.90	0.92	0.95	0.98	1.11	1.15	1.18	1.22	1.06	1.09	1.12	1.15	1.07	1.10	1.14	1.17
2017	1.13	1.17	1.22	1.27	0.93	0.97	1.01	1.05	1.15	1.19	1.24	1.29	1.13	1.17	1.22	1.27	1.10	1.14	1.19	1.24
2018	1.16	1.22	1.28	1.35	0.97	1.02	1.07	1.12	1.20	1.26	1.32	1.38	1.20	1.26	1.33	1.39	1.14	1.19	1.25	1.31
2019	1.19	1.26	1.33	1.41	1.01	1.07	1.13	1.20	1.24	1.31	1.39	1.47	1.26	1.34	1.42	1.50	1.18	1.25	1.32	1.40
2020	1.21	1.29	1.38	1.48	1.05	1.12	1.20	1.28	1.29	1.38	1.47	1.57	1.31	1.40	1.50	1.60	1.21	1.30	1.39	1.48
2021	1.23	1.33	1.44	1.55	1.09	1.18	1.27	1.37	1.33	1.44	1.56	1.68	1.35	1.46	1.58	1.70	1.24	1.34	1.45	1.57
2022	1.26	1.37	1.50	1.63	1.13	1.24	1.35	1.47	1.39	1.51	1.65	1.80	1.41	1.54	1.68	1.83	1.27	1.39	1.52	1.65
2023	1.29	1.42	1.56	1.72	1.18	1.30	1.43	1.57	1.44	1.58	1.75	1.92	1.46	1.61	1.78	1.96	1.31	1.44	1.59	1.75
2024	1.32	1.47	1.63	1.82	1.22	1.36	1.51	1.68	1.49	1.66	1.85	2.05	1.51	1.69	1.88	2.08	1.34	1.49	1.66	1.85
2025	1.35	1.51	1.70	1.91	1.27	1.43	1.60	1.80	1.55	1.74	1.95	2.19	1.56	1.75	1.97	2.20	1.38	1.55	1.74	1.95
2026	1.37	1.55	1.76	2.00	1.32	1.49	1.69	1.92	1.61	1.82	2.07	2.34	1.60	1.82	2.06	2.34	1.41	1.61	1.82	2.06
2027	1.39	1.60	1.83	2.09	1.36	1.56	1.79	2.04	1.67	1.91	2.19	2.51	1.65	1.89	2.16	2.47	1.45	1.67	1.91	2.18
2028	1.42	1.64	1.90	2.19	1.41	1.63	1.89	2.18	1.74	2.01	2.33	2.68	1.70	1.97	2.27	2.62	1.49	1.73	1.99	2.30
2029	1.45	1.69	1.98	2.31	1.46	1.71	1.99	2.32	1.80	2.11	2.46	2.86	1.75	2.05	2.39	2.78	1.53	1.79	2.09	2.44
2030	1.48	1.75	2.06	2.42	1.51	1.79	2.11	2.48	1.87	2.21	2.61	3.07	1.80	2.13	2.51	2.95	1.57	1.86	2.19	2.58
2031	1.51	1.80	2.15	2.55	1.57	1.87	2.22	2.64	1.94	2.31	2.75	3.27	1.86	2.22	2.64	3.13	1.62	1.93	2.29	2.72
2032	1.55	1.87	2.24	2.69	1.62	1.95	2.34	2.81	2.01	2.42	2.91	3.49	1.92	2.31	2.78	3.33	1.66	2.00	2.40	2.88
2033	1.59	1.93	2.34	2.84	1.68	2.04	2.48	3.00	2.09	2.54	3.08	3.73	1.99	2.42	2.93	3.55	1.70	2.06	2.51	3.03
2034	1.63	2.00	2.45	3.00	1.75	2.14	2.63	3.21	2.17	2.66	3.26	3.99	2.08	2.56	3.13	3.83	1.75	2.15	2.63	3.22
2035	1.68	2.08	2.57	3.18	1.82	2.25	2.79	3.44	2.26	2.80	3.46	4.27	2.19	2.71	3.35	4.14	1.80	2.23	2.76	3.41
2036	1.73	2.17	2.71	3.38	1.89	2.37	2.96	3.69	2.35	2.94	3.68	4.58	2.31	2.89	3.61	4.50	1.85	2.32	2.89	3.60
2037	1.79	2.26	2.86	3.59	1.97	2.49	3.14	3.95	2.44	3.09	3.90	4.90	2.44	3.08	3.88	4.88	1.91	2.41	3.04	3.82
2038	1.85	2.36	3.00	3.81	2.04	2.60	3.32	4.21	2.54	3.24	4.12	5.23	2.57	3.28	4.17	5.30	1.96	2.50	3.18	4.04
2039	1.90	2.45	3.15	4.04	2.12	2.73	3.51	4.50	2.63	3.40	4.37	5.60	2.67	3.44	4.42	5.67	2.01	2.59	3.33	4.27
2040	1.95	2.54	3.30	4.27	2.20	2.86	3.71	4.81	2.74	3.57	4.63	5.99	2.78	3.61	4.69	6.07	2.07	2.69	3.49	4.53
2041	2.00	2.63	3.45	4.51	2.28	3.00	3.93	5.14	2.85	3.75	4.91	6.42	2.89	3.80	4.98	6.50	2.12	2.79	3.66	4.78
2042	2.05	2.72	3.60	4.75	2.37	3.15	4.17	5.50	2.97	3.95	5.22	6.89	3.00	3.99	5.28	6.96	2.18	2.90	3.83	5.06
2043	2.10	2.81	3.76	5.01	2.47	3.31	4.42	5.89	3.10	4.15	5.55	7.39	3.12	4.19	5.59	7.46	2.25	3.01	4.02	5.36

54

Table S-2, continued. Projected fuel price indices with assumed general price inflation rates of 2 %, 3 %, 4 %, and 5 %, by end-use sector and fuel type.

Census Region 2 (Illinois, Indiana, Iowa, Kansas, Michigan, Minnesota, Missouri, Nebraska, North Dakota, Ohio, South Dakota, Wisconsin)

Projected April 1 Fuel Price Indices (April 1, 2013 = 1.00)

--------INDUSTRIAL--------

Year	Electricity				Distillate Oil				Residual Oil				Natural Gas				Coal			
	Inflation Rate				Inflation Rate				Inflation Rate				Inflation Rate				Inflation Rate			
	2 %	3 %	4 %	5 %	2 %	3 %	4 %	5 %	2 %	3 %	4 %	5 %	2 %	3 %	4 %	5 %	2 %	3 %	4 %	5 %
2014	1.01	1.02	1.03	1.04	0.89	0.90	0.91	0.92	1.10	1.11	1.12	1.13	1.04	1.05	1.06	1.07	1.02	1.03	1.04	1.05
2015	1.02	1.04	1.06	1.08	0.88	0.89	0.91	0.93	1.13	1.16	1.18	1.20	1.06	1.09	1.11	1.13	1.06	1.08	1.10	1.12
2016	1.06	1.09	1.12	1.16	0.91	0.93	0.96	0.99	1.16	1.20	1.23	1.27	1.16	1.19	1.23	1.27	1.09	1.12	1.16	1.19
2017	1.11	1.16	1.20	1.25	0.94	0.98	1.02	1.06	1.20	1.25	1.30	1.35	1.25	1.30	1.35	1.40	1.12	1.17	1.21	1.26
2018	1.15	1.21	1.27	1.33	0.98	1.03	1.08	1.13	1.25	1.31	1.37	1.44	1.35	1.41	1.48	1.56	1.15	1.21	1.27	1.33
2019	1.17	1.24	1.32	1.39	1.02	1.08	1.15	1.22	1.29	1.37	1.45	1.54	1.42	1.51	1.60	1.69	1.19	1.26	1.34	1.42
2020	1.18	1.27	1.36	1.45	1.06	1.13	1.21	1.30	1.34	1.43	1.54	1.64	1.48	1.58	1.69	1.81	1.22	1.31	1.40	1.50
2021	1.21	1.31	1.41	1.53	1.10	1.19	1.28	1.39	1.39	1.50	1.62	1.75	1.53	1.66	1.79	1.93	1.26	1.36	1.47	1.58
2022	1.24	1.35	1.48	1.61	1.15	1.25	1.37	1.49	1.44	1.58	1.72	1.87	1.60	1.75	1.91	2.08	1.29	1.41	1.54	1.68
2023	1.28	1.41	1.55	1.71	1.19	1.31	1.45	1.59	1.50	1.65	1.82	2.00	1.67	1.85	2.03	2.24	1.33	1.47	1.62	1.78
2024	1.31	1.46	1.62	1.81	1.23	1.37	1.53	1.70	1.55	1.73	1.92	2.14	1.74	1.93	2.15	2.39	1.37	1.53	1.70	1.89
2025	1.34	1.51	1.70	1.90	1.28	1.44	1.62	1.81	1.61	1.81	2.03	2.28	1.79	2.01	2.26	2.53	1.41	1.58	1.78	1.99
2026	1.37	1.56	1.77	2.00	1.33	1.50	1.71	1.93	1.67	1.90	2.15	2.44	1.84	2.09	2.37	2.69	1.45	1.65	1.87	2.11
2027	1.40	1.61	1.84	2.10	1.37	1.57	1.80	2.06	1.74	1.99	2.28	2.61	1.90	2.18	2.49	2.85	1.49	1.71	1.96	2.24
2028	1.43	1.66	1.92	2.21	1.42	1.64	1.90	2.19	1.81	2.09	2.42	2.79	1.96	2.27	2.62	3.03	1.54	1.78	2.06	2.38
2029	1.47	1.72	2.00	2.33	1.47	1.71	2.00	2.33	1.87	2.19	2.56	2.98	2.03	2.37	2.76	3.22	1.58	1.85	2.16	2.52
2030	1.50	1.77	2.09	2.46	1.51	1.79	2.11	2.48	1.95	2.30	2.71	3.19	2.09	2.46	2.90	3.41	1.64	1.93	2.27	2.68
2031	1.54	1.84	2.18	2.59	1.56	1.86	2.22	2.63	2.02	2.40	2.86	3.40	2.16	2.57	3.06	3.63	1.68	2.00	2.38	2.83
2032	1.58	1.91	2.29	2.75	1.62	1.95	2.34	2.80	2.09	2.52	3.02	3.63	2.24	2.69	3.23	3.88	1.73	2.08	2.50	3.00
2033	1.63	1.98	2.40	2.90	1.68	2.04	2.47	2.99	2.17	2.64	3.20	3.87	2.32	2.83	3.43	4.15	1.77	2.15	2.61	3.17
2034	1.68	2.06	2.52	3.08	1.74	2.14	2.62	3.20	2.25	2.76	3.39	4.14	2.45	3.00	3.68	4.50	1.82	2.24	2.74	3.35
2035	1.73	2.15	2.65	3.28	1.81	2.24	2.77	3.42	2.35	2.91	3.60	4.44	2.59	3.21	3.97	4.90	1.88	2.33	2.89	3.56
2036	1.80	2.25	2.81	3.51	1.88	2.36	2.95	3.67	2.44	3.06	3.82	4.76	2.76	3.45	4.31	5.37	1.94	2.42	3.03	3.77
2037	1.87	2.36	2.98	3.75	1.96	2.48	3.12	3.93	2.54	3.21	4.04	5.09	2.93	3.71	4.67	5.88	1.99	2.52	3.18	4.00
2038	1.94	2.47	3.15	4.00	2.03	2.59	3.30	4.19	2.63	3.36	4.28	5.43	3.11	3.97	5.06	6.43	2.06	2.62	3.34	4.25
2039	2.00	2.58	3.31	4.25	2.11	2.71	3.49	4.47	2.73	3.52	4.53	5.81	3.25	4.19	5.39	6.91	2.12	2.73	3.51	4.50
2040	2.06	2.68	3.48	4.51	2.19	2.84	3.69	4.78	2.84	3.70	4.80	6.22	3.40	4.42	5.74	7.43	2.18	2.83	3.68	4.76
2041	2.12	2.79	3.65	4.78	2.27	2.98	3.91	5.11	2.96	3.89	5.10	6.66	3.55	4.66	6.11	7.99	2.24	2.95	3.87	5.05
2042	2.18	2.89	3.82	5.05	2.36	3.13	4.14	5.46	3.08	4.09	5.41	7.14	3.71	4.92	6.52	8.60	2.31	3.06	4.05	5.35
2043	2.24	3.00	4.00	5.34	2.45	3.28	4.38	5.84	3.21	4.30	5.75	7.66	3.88	5.20	6.94	9.25	2.37	3.18	4.25	5.66

Table S-3. Projected fuel price indices with assumed general price inflation rates of 2 %, 3 %, 4 %, and 5 %, by end-use sector and fuel type.

Census Region 3 (Alabama, Arkansas, Delaware, District of Columbia, Florida, Georgia, Kentucky, Louisiana, Maryland, Mississippi, North Carolina, Oklahoma, South Carolina, Tennessee, Texas, Virginia, West Virginia)

Projected April 1 Fuel Price Indices (April 1, 2013 = 1.00)

—RESIDENTIAL—

Year	Electricity				Distillate Oil				LPG				Natural Gas			
	Inflation Rate				Inflation Rate				Inflation Rate				Inflation Rate			
	2 %	3 %	4 %	5 %	2 %	3 %	4 %	5 %	2 %	3 %	4 %	5 %	2 %	3 %	4 %	5 %
2014	1.03	1.04	1.05	1.06	0.99	1.00	1.01	1.02	0.99	1.00	1.01	1.02	1.00	1.01	1.02	1.03
2015	1.06	1.08	1.10	1.13	1.00	1.02	1.04	1.06	0.98	1.00	1.02	1.04	1.00	1.02	1.04	1.06
2016	1.10	1.13	1.17	1.20	1.03	1.07	1.10	1.13	1.00	1.03	1.06	1.09	1.05	1.08	1.11	1.14
2017	1.13	1.18	1.23	1.27	1.07	1.12	1.16	1.21	1.04	1.09	1.13	1.17	1.10	1.14	1.19	1.23
2018	1.16	1.22	1.28	1.34	1.11	1.17	1.22	1.28	1.09	1.15	1.20	1.26	1.15	1.21	1.27	1.33
2019	1.18	1.25	1.33	1.41	1.15	1.22	1.29	1.37	1.13	1.20	1.27	1.35	1.19	1.26	1.34	1.42
2020	1.21	1.29	1.38	1.48	1.19	1.28	1.36	1.46	1.18	1.26	1.35	1.44	1.23	1.32	1.41	1.51
2021	1.24	1.34	1.45	1.56	1.23	1.33	1.44	1.56	1.22	1.31	1.42	1.53	1.27	1.37	1.48	1.60
2022	1.27	1.39	1.51	1.65	1.28	1.39	1.52	1.66	1.26	1.37	1.50	1.63	1.32	1.44	1.57	1.71
2023	1.29	1.43	1.57	1.73	1.32	1.46	1.61	1.77	1.30	1.43	1.57	1.73	1.38	1.52	1.67	1.84
2024	1.32	1.47	1.64	1.82	1.37	1.53	1.70	1.89	1.34	1.49	1.65	1.84	1.42	1.58	1.76	1.96
2025	1.35	1.52	1.71	1.91	1.42	1.60	1.79	2.01	1.38	1.55	1.74	1.95	1.47	1.65	1.85	2.08
2026	1.39	1.57	1.78	2.02	1.47	1.67	1.89	2.15	1.42	1.61	1.82	2.06	1.51	1.72	1.95	2.21
2027	1.42	1.63	1.86	2.13	1.52	1.75	2.00	2.29	1.46	1.67	1.91	2.19	1.56	1.79	2.05	2.34
2028	1.45	1.68	1.95	2.25	1.58	1.83	2.11	2.44	1.50	1.73	2.00	2.31	1.61	1.86	2.15	2.49
2029	1.49	1.74	2.03	2.37	1.63	1.91	2.23	2.60	1.54	1.80	2.10	2.44	1.66	1.94	2.27	2.64
2030	1.52	1.80	2.12	2.49	1.69	1.99	2.35	2.77	1.58	1.86	2.19	2.58	1.72	2.03	2.39	2.81
2031	1.55	1.85	2.20	2.62	1.75	2.08	2.48	2.94	1.62	1.93	2.30	2.73	1.77	2.11	2.52	2.99
2032	1.59	1.91	2.30	2.76	1.81	2.17	2.61	3.13	1.66	2.00	2.40	2.88	1.83	2.20	2.65	3.17
2033	1.63	1.98	2.40	2.90	1.87	2.27	2.76	3.34	1.71	2.07	2.52	3.05	1.89	2.30	2.79	3.38
2034	1.67	2.05	2.51	3.07	1.94	2.38	2.91	3.56	1.75	2.15	2.64	3.22	1.97	2.42	2.96	3.62
2035	1.72	2.14	2.64	3.26	2.01	2.49	3.08	3.80	1.81	2.24	2.77	3.42	2.05	2.55	3.15	3.89
2036	1.78	2.22	2.78	3.46	2.09	2.61	3.26	4.06	1.86	2.33	2.91	3.62	2.15	2.70	3.37	4.19
2037	1.83	2.32	2.92	3.68	2.17	2.74	3.45	4.35	1.92	2.42	3.05	3.84	2.25	2.85	3.59	4.52
2038	1.89	2.42	3.07	3.91	2.24	2.86	3.64	4.63	1.97	2.52	3.21	4.07	2.36	3.01	3.83	4.87
2039	1.95	2.51	3.23	4.15	2.32	2.99	3.85	4.94	2.03	2.62	3.37	4.32	2.44	3.14	4.04	5.18
2040	2.00	2.61	3.38	4.38	2.41	3.14	4.07	5.27	2.09	2.72	3.53	4.57	2.53	3.29	4.28	5.54
2041	2.05	2.70	3.54	4.62	2.50	3.28	4.30	5.63	2.15	2.83	3.71	4.84	2.62	3.44	4.51	5.90
2042	2.11	2.79	3.70	4.88	2.59	3.44	4.55	6.01	2.22	2.94	3.89	5.14	2.71	3.60	4.76	6.28
2043	2.16	2.89	3.86	5.15	2.69	3.61	4.82	6.42	2.28	3.06	4.09	5.45	2.80	3.76	5.02	6.69

56

Table S-3, continued. Projected fuel price indices with assumed general price inflation rates of 2 %, 3 %, 4 %, and 5 %, by end-use sector and fuel type.

Census Region 3 (Alabama, Arkansas, Delaware, District of Columbia, Florida, Georgia, Kentucky, Louisiana, Maryland, Mississippi, North Carolina, Oklahoma, South Carolina, Tennessee, Texas, Virginia, West Virginia)

Projected April 1 Fuel Price Indices (April 1, 2013 = 1.00)

-------------------------------------COMMERCIAL-------------------------------------

Year	Electricity				Distillate Oil				Residual Oil				Natural Gas				Coal			
	Inflation Rate				Inflation Rate				Inflation Rate				Inflation Rate				Inflation Rate			
	2 %	3 %	4 %	5 %	2 %	3 %	4 %	5 %	2 %	3 %	4 %	5 %	2 %	3 %	4 %	5 %	2 %	3 %	4 %	5 %
2014	1.02	1.03	1.04	1.05	0.90	0.91	0.92	0.93	0.86	0.87	0.88	0.89	1.01	1.02	1.03	1.04	1.01	1.02	1.03	1.04
2015	1.05	1.07	1.09	1.11	0.89	0.91	0.93	0.94	0.83	0.85	0.87	0.88	1.00	1.02	1.04	1.06	1.05	1.07	1.09	1.11
2016	1.08	1.12	1.15	1.18	0.92	0.95	0.98	1.00	0.86	0.89	0.91	0.94	1.06	1.09	1.12	1.16	1.07	1.10	1.13	1.17
2017	1.12	1.16	1.21	1.26	0.96	1.00	1.04	1.08	0.89	0.93	0.96	1.00	1.12	1.16	1.21	1.25	1.09	1.13	1.18	1.22
2018	1.15	1.21	1.27	1.33	1.00	1.05	1.10	1.15	0.93	0.98	1.03	1.08	1.17	1.23	1.29	1.36	1.11	1.16	1.22	1.28
2019	1.17	1.24	1.31	1.39	1.04	1.10	1.16	1.23	0.97	1.02	1.09	1.15	1.22	1.29	1.37	1.45	1.12	1.19	1.26	1.33
2020	1.19	1.28	1.37	1.46	1.07	1.15	1.23	1.31	1.01	1.08	1.15	1.23	1.26	1.35	1.44	1.54	1.14	1.22	1.30	1.39
2021	1.22	1.32	1.43	1.54	1.12	1.21	1.30	1.41	1.05	1.13	1.22	1.32	1.30	1.41	1.52	1.64	1.16	1.25	1.35	1.46
2022	1.26	1.37	1.50	1.63	1.16	1.27	1.38	1.51	1.09	1.19	1.30	1.42	1.36	1.48	1.62	1.76	1.18	1.29	1.40	1.53
2023	1.28	1.41	1.55	1.71	1.21	1.33	1.46	1.61	1.13	1.25	1.38	1.52	1.42	1.56	1.72	1.90	1.21	1.33	1.46	1.61
2024	1.30	1.45	1.61	1.79	1.25	1.39	1.55	1.72	1.18	1.32	1.46	1.63	1.47	1.64	1.82	2.02	1.24	1.38	1.54	1.71
2025	1.32	1.49	1.67	1.88	1.30	1.46	1.64	1.84	1.23	1.38	1.55	1.74	1.52	1.70	1.91	2.15	1.27	1.43	1.61	1.80
2026	1.35	1.53	1.74	1.97	1.34	1.53	1.73	1.96	1.28	1.45	1.65	1.86	1.57	1.78	2.02	2.28	1.30	1.48	1.68	1.90
2027	1.38	1.58	1.81	2.07	1.39	1.60	1.83	2.09	1.33	1.53	1.75	2.00	1.61	1.85	2.11	2.42	1.34	1.53	1.75	2.01
2028	1.42	1.64	1.89	2.19	1.44	1.67	1.93	2.23	1.39	1.61	1.86	2.15	1.66	1.92	2.22	2.57	1.37	1.59	1.84	2.12
2029	1.45	1.70	1.98	2.31	1.49	1.75	2.04	2.38	1.44	1.69	1.97	2.30	1.71	2.00	2.34	2.73	1.41	1.65	1.92	2.24
2030	1.49	1.76	2.07	2.43	1.54	1.82	2.15	2.53	1.51	1.78	2.10	2.47	1.77	2.09	2.46	2.89	1.45	1.71	2.02	2.38
2031	1.52	1.81	2.16	2.56	1.60	1.90	2.27	2.69	1.55	1.85	2.20	2.61	1.83	2.18	2.59	3.08	1.49	1.77	2.11	2.51
2032	1.56	1.88	2.25	2.70	1.65	1.99	2.39	2.87	1.62	1.95	2.34	2.80	1.88	2.26	2.72	3.26	1.53	1.84	2.22	2.66
2033	1.60	1.94	2.36	2.85	1.71	2.08	2.53	3.06	1.68	2.05	2.48	3.01	1.94	2.36	2.86	3.47	1.57	1.91	2.32	2.81
2034	1.65	2.02	2.48	3.03	1.78	2.18	2.68	3.27	1.74	2.14	2.62	3.21	2.03	2.49	3.05	3.73	1.61	1.98	2.42	2.96
2035	1.70	2.11	2.61	3.22	1.85	2.29	2.84	3.50	1.83	2.26	2.80	3.46	2.12	2.63	3.25	4.01	1.66	2.06	2.55	3.15
2036	1.77	2.21	2.76	3.44	1.93	2.41	3.01	3.75	1.91	2.38	2.98	3.71	2.23	2.79	3.49	4.34	1.71	2.14	2.68	3.34
2037	1.83	2.32	2.92	3.67	2.00	2.53	3.19	4.02	1.99	2.51	3.17	3.99	2.34	2.96	3.74	4.70	1.76	2.22	2.80	3.53
2038	1.90	2.43	3.09	3.92	2.07	2.65	3.37	4.28	2.06	2.63	3.34	4.25	2.46	3.14	4.00	5.08	1.81	2.31	2.94	3.74
2039	1.97	2.54	3.26	4.18	2.15	2.78	3.57	4.58	2.15	2.77	3.56	4.57	2.55	3.28	4.26	5.41	1.86	2.40	3.08	3.95
2040	2.03	2.64	3.43	4.44	2.24	2.91	3.78	4.89	2.24	2.92	3.79	4.90	2.64	3.44	4.46	5.78	1.92	2.49	3.24	4.19
2041	2.08	2.74	3.59	4.69	2.32	3.05	4.00	5.23	2.34	3.07	4.03	5.26	2.75	3.61	4.73	6.18	1.97	2.59	3.39	4.43
2042	2.14	2.84	3.75	4.95	2.41	3.20	4.24	5.60	2.44	3.24	4.29	5.66	2.85	3.78	5.01	6.61	2.03	2.69	3.56	4.69
2043	2.19	2.94	3.93	5.23	2.51	3.36	4.49	5.99	2.55	3.42	4.57	6.09	2.96	3.97	5.30	7.06	2.08	2.79	3.72	4.96

Table S-3, continued. Projected fuel price indices with assumed general price inflation rates of 2 %, 3 %, 4 %, and 5 %, by end-use sector and fuel type.

Census Region 3 (Alabama, Arkansas, Delaware, District of Columbia, Florida, Georgia, Kentucky, Louisiana, Maryland, Mississippi, North Carolina, Oklahoma, South Carolina, Tennessee, Texas, Virginia, West Virginia)

Projected April 1 Fuel Price Indices (April 1, 2013 = 1.00)

--------INDUSTRIAL--------

Year	Electricity				Distillate Oil				Residual Oil				Natural Gas				Coal			
	Inflation Rate				Inflation Rate				Inflation Rate				Inflation Rate				Inflation Rate			
	2 %	3 %	4 %	5 %	2 %	3 %	4 %	5 %	2 %	3 %	4 %	5 %	2 %	3 %	4 %	5 %	2 %	3 %	4 %	5 %
2014	1.02	1.03	1.04	1.05	0.92	0.93	0.94	0.95	0.89	0.90	0.91	0.92	1.08	1.09	1.10	1.12	1.01	1.02	1.03	1.04
2015	1.04	1.06	1.08	1.10	0.91	0.93	0.95	0.97	0.87	0.89	0.91	0.93	1.12	1.15	1.17	1.19	1.05	1.07	1.09	1.11
2016	1.08	1.11	1.15	1.18	0.94	0.97	1.00	1.03	0.90	0.93	0.95	0.98	1.26	1.29	1.33	1.37	1.08	1.12	1.15	1.18
2017	1.12	1.17	1.21	1.26	0.98	1.02	1.06	1.10	0.93	0.97	1.01	1.04	1.35	1.41	1.46	1.52	1.11	1.15	1.20	1.25
2018	1.15	1.21	1.27	1.33	1.02	1.07	1.13	1.18	0.97	1.02	1.07	1.12	1.44	1.51	1.59	1.67	1.14	1.19	1.25	1.31
2019	1.18	1.25	1.33	1.40	1.06	1.13	1.20	1.27	1.01	1.07	1.13	1.20	1.51	1.60	1.70	1.80	1.16	1.23	1.30	1.38
2020	1.21	1.29	1.38	1.48	1.10	1.18	1.26	1.35	1.05	1.12	1.20	1.29	1.56	1.68	1.79	1.92	1.19	1.27	1.36	1.45
2021	1.24	1.34	1.45	1.56	1.15	1.24	1.34	1.44	1.09	1.18	1.28	1.38	1.63	1.76	1.90	2.05	1.21	1.31	1.42	1.53
2022	1.28	1.39	1.52	1.66	1.19	1.30	1.42	1.55	1.14	1.24	1.35	1.47	1.72	1.88	2.05	2.23	1.24	1.36	1.48	1.62
2023	1.31	1.44	1.59	1.75	1.24	1.37	1.50	1.66	1.18	1.30	1.43	1.58	1.83	2.01	2.22	2.44	1.28	1.41	1.55	1.70
2024	1.34	1.49	1.66	1.84	1.28	1.43	1.59	1.77	1.23	1.37	1.52	1.69	1.91	2.13	2.37	2.63	1.31	1.46	1.62	1.80
2025	1.37	1.54	1.73	1.94	1.33	1.50	1.68	1.88	1.28	1.44	1.61	1.81	1.98	2.23	2.50	2.80	1.34	1.51	1.69	1.90
2026	1.41	1.60	1.81	2.05	1.38	1.56	1.77	2.01	1.33	1.51	1.71	1.94	2.07	2.34	2.66	3.01	1.38	1.56	1.77	2.01
2027	1.44	1.66	1.90	2.17	1.42	1.63	1.87	2.14	1.39	1.59	1.82	2.08	2.14	2.45	2.81	3.21	1.41	1.62	1.85	2.12
2028	1.49	1.72	1.99	2.30	1.47	1.70	1.97	2.28	1.45	1.67	1.93	2.23	2.22	2.57	2.97	3.42	1.45	1.67	1.94	2.23
2029	1.53	1.79	2.09	2.44	1.52	1.78	2.08	2.42	1.50	1.75	2.05	2.39	2.30	2.69	3.14	3.66	1.49	1.74	2.03	2.36
2030	1.57	1.86	2.19	2.58	1.57	1.86	2.19	2.57	1.56	1.85	2.18	2.56	2.38	2.81	3.31	3.90	1.53	1.80	2.12	2.50
2031	1.62	1.93	2.29	2.72	1.62	1.93	2.30	2.74	1.61	1.92	2.29	2.72	2.48	2.95	3.51	4.17	1.57	1.87	2.22	2.64
2032	1.66	2.00	2.40	2.88	1.68	2.02	2.43	2.91	1.68	2.02	2.43	2.92	2.55	3.08	3.69	4.43	1.61	1.94	2.33	2.79
2033	1.71	2.08	2.52	3.05	1.74	2.11	2.56	3.11	1.75	2.13	2.58	3.12	2.66	3.23	3.92	4.75	1.65	2.01	2.44	2.95
2034	1.77	2.17	2.66	3.25	1.81	2.22	2.71	3.32	1.82	2.23	2.73	3.34	2.82	3.46	4.23	5.18	1.70	2.09	2.56	3.12
2035	1.84	2.28	2.82	3.48	1.88	2.33	2.88	3.55	1.90	2.35	2.91	3.59	2.99	3.71	4.59	5.66	1.75	2.17	2.68	3.31
2036	1.92	2.40	3.00	3.74	1.95	2.44	3.05	3.80	1.98	2.48	3.09	3.86	3.21	4.02	5.02	6.26	1.80	2.25	2.81	3.51
2037	2.00	2.53	3.19	4.01	2.03	2.57	3.23	4.07	2.06	2.61	3.29	4.14	3.45	4.36	5.50	6.92	1.85	2.34	2.95	3.72
2038	2.08	2.66	3.38	4.30	2.10	2.68	3.41	4.34	2.14	2.73	3.48	4.41	3.69	4.70	5.99	7.61	1.91	2.43	3.10	3.94
2039	2.16	2.79	3.58	4.60	2.18	2.81	3.61	4.63	2.23	2.87	3.69	4.74	3.84	4.95	6.37	8.17	1.96	2.52	3.25	4.16
2040	2.24	2.91	3.78	4.89	2.26	2.95	3.82	4.95	2.32	3.02	3.93	5.08	4.03	5.24	6.80	8.81	2.02	2.62	3.40	4.41
2041	2.30	3.03	3.97	5.19	2.35	3.09	4.05	5.29	2.42	3.19	4.18	5.46	4.23	5.56	7.29	9.53	2.07	2.72	3.56	4.66
2042	2.37	3.15	4.17	5.50	2.44	3.24	4.29	5.66	2.53	3.36	4.44	5.87	4.45	5.90	7.81	10.31	2.12	2.82	3.73	4.92
2043	2.45	3.28	4.38	5.84	2.54	3.40	4.54	6.05	2.64	3.54	4.73	6.31	4.67	6.26	8.36	11.14	2.19	2.93	3.91	5.22

Table S-4. Projected fuel price indices with assumed general price inflation rates of 2 %, 3 %, 4 %, and 5 %, by end-use sector and fuel type.

Census Region 4 (Alaska, Arizona, California, Colorado, Hawaii, Idaho, Montana, Nevada, New Mexico, Oregon, Utah, Washington, Wyoming)

Projected April 1 Fuel Price Indices (April 1, 2013 = 1.00)

-----------------RESIDENTIAL-----------------

Year	Electricity				Distillate Oil				LPG				Natural Gas			
	Inflation Rate				Inflation Rate				Inflation Rate				Inflation Rate			
	2 %	3 %	4 %	5 %	2 %	3 %	4 %	5 %	2 %	3 %	4 %	5 %	2 %	3 %	4 %	5 %
2014	1.02	1.03	1.04	1.05	0.93	0.94	0.95	0.96	0.99	1.00	1.01	1.02	1.02	1.03	1.04	1.05
2015	1.04	1.06	1.08	1.10	0.93	0.95	0.97	0.99	0.98	1.00	1.02	1.04	1.07	1.09	1.11	1.13
2016	1.07	1.10	1.13	1.16	0.96	0.99	1.02	1.05	1.00	1.03	1.06	1.09	1.15	1.18	1.21	1.25
2017	1.09	1.14	1.18	1.23	1.00	1.04	1.08	1.12	1.04	1.09	1.13	1.17	1.22	1.27	1.32	1.37
2018	1.11	1.17	1.23	1.29	1.04	1.09	1.14	1.20	1.09	1.14	1.20	1.26	1.30	1.36	1.43	1.50
2019	1.13	1.20	1.27	1.34	1.08	1.14	1.21	1.28	1.13	1.20	1.27	1.35	1.37	1.45	1.54	1.63
2020	1.15	1.23	1.32	1.41	1.11	1.19	1.28	1.36	1.17	1.26	1.35	1.44	1.42	1.52	1.63	1.74
2021	1.17	1.27	1.37	1.48	1.15	1.25	1.35	1.46	1.21	1.31	1.42	1.53	1.48	1.60	1.73	1.87
2022	1.19	1.30	1.41	1.54	1.20	1.31	1.42	1.55	1.25	1.37	1.49	1.63	1.54	1.68	1.84	2.00
2023	1.20	1.33	1.46	1.61	1.24	1.37	1.51	1.66	1.29	1.43	1.57	1.73	1.60	1.77	1.95	2.14
2024	1.23	1.37	1.52	1.69	1.28	1.43	1.59	1.77	1.33	1.48	1.65	1.83	1.66	1.84	2.05	2.28
2025	1.26	1.41	1.59	1.78	1.33	1.50	1.68	1.89	1.37	1.54	1.73	1.94	1.70	1.91	2.14	2.41
2026	1.29	1.46	1.65	1.87	1.38	1.57	1.78	2.01	1.41	1.60	1.82	2.06	1.74	1.98	2.24	2.54
2027	1.31	1.50	1.72	1.97	1.43	1.64	1.88	2.15	1.45	1.66	1.90	2.18	1.79	2.05	2.34	2.68
2028	1.34	1.55	1.79	2.06	1.48	1.72	1.98	2.29	1.49	1.72	1.99	2.30	1.84	2.13	2.46	2.84
2029	1.37	1.60	1.87	2.18	1.54	1.80	2.10	2.44	1.53	1.79	2.09	2.43	1.90	2.22	2.59	3.01
2030	1.40	1.66	1.95	2.30	1.59	1.88	2.21	2.61	1.57	1.85	2.18	2.57	1.95	2.31	2.72	3.20
2031	1.45	1.73	2.05	2.44	1.65	1.96	2.34	2.77	1.61	1.92	2.29	2.72	2.02	2.40	2.86	3.40
2032	1.49	1.80	2.16	2.59	1.70	2.05	2.46	2.96	1.65	1.99	2.39	2.87	2.09	2.51	3.02	3.62
2033	1.53	1.86	2.25	2.73	1.76	2.14	2.60	3.15	1.70	2.06	2.50	3.03	2.16	2.63	3.19	3.86
2034	1.56	1.92	2.35	2.87	1.83	2.24	2.75	3.36	1.74	2.14	2.62	3.21	2.25	2.76	3.38	4.14
2035	1.61	2.00	2.47	3.05	1.90	2.35	2.91	3.59	1.79	2.22	2.75	3.39	2.35	2.92	3.61	4.45
2036	1.66	2.08	2.60	3.24	1.97	2.47	3.08	3.84	1.85	2.31	2.89	3.60	2.47	3.09	3.86	4.81
2037	1.72	2.17	2.74	3.45	2.05	2.59	3.27	4.11	1.90	2.40	3.03	3.81	2.59	3.27	4.12	5.19
2038	1.77	2.27	2.88	3.66	2.12	2.70	3.44	4.37	1.96	2.50	3.18	4.04	2.72	3.47	4.41	5.61
2039	1.83	2.36	3.03	3.89	2.20	2.83	3.64	4.67	2.02	2.60	3.34	4.28	2.82	3.64	4.68	6.00
2040	1.89	2.46	3.20	4.14	2.27	2.96	3.84	4.98	2.07	2.70	3.50	4.53	2.93	3.81	4.95	6.41
2041	1.94	2.55	3.35	4.37	2.36	3.10	4.07	5.31	2.13	2.80	3.67	4.80	3.03	3.99	5.23	6.83
2042	1.99	2.64	3.49	4.61	2.45	3.25	4.30	5.68	2.20	2.92	3.86	5.09	3.14	4.17	5.52	7.28
2043	2.04	2.73	3.65	4.86	2.54	3.41	4.56	6.07	2.26	3.03	4.05	5.40	3.25	4.36	5.82	7.76

59

Table S-4, continued. Projected fuel price indices with assumed general price inflation rates of 2 %, 3 %, 4 %, and 5 %, by end-use sector and fuel type.

Census Region 4 (Alaska, Arizona, California, Colorado, Hawaii, Idaho, Montana, Nevada, New Mexico, Oregon, Utah, Washington, Wyoming)

Projected April 1 Fuel Price Indices (April 1, 2013 = 1.00)

-----COMMERCIAL-----

Year	Electricity				Distillate Oil				Residual Oil				Natural Gas				Coal			
	Inflation Rate				Inflation Rate				Inflation Rate				Inflation Rate				Inflation Rate			
	2 %	3 %	4 %	5 %	2 %	3 %	4 %	5 %	2 %	3 %	4 %	5 %	2 %	3 %	4 %	5 %	2 %	3 %	4 %	5 %
2014	1.02	1.03	1.04	1.05	0.89	0.90	0.91	0.92	0.76	0.77	0.77	0.78	1.03	1.04	1.05	1.06	1.03	1.04	1.05	1.06
2015	1.04	1.06	1.08	1.10	0.88	0.90	0.91	0.93	0.70	0.71	0.73	0.74	1.05	1.07	1.09	1.12	1.07	1.09	1.11	1.13
2016	1.07	1.10	1.14	1.17	0.91	0.94	0.97	0.99	0.72	0.74	0.76	0.78	1.13	1.16	1.20	1.23	1.11	1.14	1.18	1.21
2017	1.10	1.14	1.19	1.23	0.95	0.99	1.03	1.07	0.75	0.78	0.81	0.84	1.21	1.25	1.30	1.35	1.14	1.19	1.23	1.28
2018	1.12	1.18	1.23	1.30	0.99	1.04	1.09	1.15	0.78	0.82	0.86	0.90	1.29	1.35	1.42	1.49	1.17	1.23	1.29	1.35
2019	1.14	1.20	1.28	1.35	1.03	1.09	1.16	1.23	0.81	0.86	0.91	0.97	1.36	1.44	1.53	1.62	1.20	1.27	1.35	1.43
2020	1.16	1.24	1.33	1.42	1.07	1.14	1.22	1.31	0.84	0.90	0.97	1.03	1.42	1.52	1.62	1.73	1.24	1.32	1.42	1.51
2021	1.19	1.29	1.39	1.50	1.11	1.20	1.30	1.40	0.88	0.95	1.03	1.11	1.48	1.60	1.73	1.86	1.27	1.38	1.49	1.61
2022	1.21	1.32	1.44	1.57	1.16	1.26	1.38	1.50	0.91	1.00	1.09	1.19	1.54	1.69	1.84	2.00	1.31	1.43	1.56	1.70
2023	1.22	1.34	1.48	1.63	1.20	1.32	1.46	1.60	0.96	1.06	1.16	1.28	1.61	1.77	1.95	2.15	1.35	1.49	1.64	1.80
2024	1.24	1.38	1.53	1.70	1.24	1.38	1.54	1.71	1.01	1.12	1.25	1.39	1.66	1.85	2.06	2.29	1.39	1.55	1.72	1.92
2025	1.26	1.42	1.60	1.79	1.29	1.45	1.63	1.83	1.05	1.18	1.33	1.49	1.70	1.92	2.15	2.41	1.43	1.61	1.81	2.03
2026	1.30	1.47	1.67	1.89	1.34	1.52	1.72	1.95	1.09	1.24	1.41	1.59	1.75	1.98	2.25	2.55	1.48	1.68	1.90	2.16
2027	1.32	1.51	1.73	1.98	1.38	1.59	1.82	2.08	1.14	1.30	1.49	1.71	1.79	2.05	2.35	2.69	1.54	1.77	2.02	2.31
2028	1.33	1.54	1.78	2.06	1.43	1.66	1.92	2.22	1.18	1.37	1.58	1.83	1.84	2.13	2.46	2.84	1.59	1.84	2.13	2.46
2029	1.36	1.58	1.85	2.16	1.49	1.74	2.03	2.36	1.23	1.44	1.68	1.96	1.90	2.22	2.59	3.02	1.64	1.92	2.24	2.61
2030	1.39	1.64	1.93	2.27	1.54	1.82	2.14	2.52	1.28	1.51	1.79	2.10	1.95	2.30	2.72	3.20	1.71	2.02	2.38	2.80
2031	1.43	1.70	2.02	2.40	1.59	1.89	2.25	2.68	1.35	1.60	1.91	2.27	2.01	2.40	2.86	3.39	1.77	2.11	2.51	2.98
2032	1.47	1.77	2.12	2.54	1.64	1.98	2.38	2.85	1.41	1.69	2.03	2.44	2.08	2.51	3.01	3.62	1.83	2.20	2.65	3.18
2033	1.50	1.82	2.21	2.68	1.70	2.07	2.51	3.04	1.46	1.78	2.16	2.61	2.16	2.63	3.19	3.86	1.90	2.31	2.80	3.40
2034	1.54	1.89	2.31	2.83	1.77	2.17	2.66	3.25	1.52	1.87	2.29	2.80	2.25	2.77	3.39	4.14	1.97	2.42	2.96	3.62
2035	1.59	1.97	2.43	3.00	1.84	2.28	2.82	3.48	1.59	1.97	2.44	3.01	2.36	2.93	3.62	4.47	2.04	2.52	3.12	3.85
2036	1.64	2.05	2.57	3.20	1.91	2.40	2.99	3.73	1.65	2.06	2.58	3.21	2.48	3.11	3.88	4.83	2.12	2.65	3.31	4.12
2037	1.70	2.14	2.70	3.40	1.99	2.52	3.17	3.99	1.71	2.17	2.73	3.44	2.61	3.30	4.16	5.24	2.19	2.77	3.49	4.39
2038	1.76	2.24	2.85	3.62	2.06	2.63	3.35	4.25	1.80	2.29	2.92	3.71	2.75	3.51	4.47	5.67	2.26	2.88	3.67	4.66
2039	1.82	2.34	3.01	3.86	2.14	2.75	3.54	4.54	1.88	2.42	3.12	4.00	2.86	3.69	4.74	6.08	2.35	3.03	3.89	4.99
2040	1.88	2.44	3.17	4.11	2.22	2.88	3.74	4.85	1.96	2.55	3.31	4.29	2.98	3.87	5.03	6.51	2.43	3.16	4.10	5.31
2041	1.93	2.53	3.32	4.34	2.30	3.02	3.96	5.18	2.05	2.69	3.52	4.61	3.09	4.06	5.32	6.95	2.49	3.28	4.29	5.61
2042	1.97	2.62	3.46	4.57	2.39	3.17	4.20	5.54	2.14	2.84	3.75	4.95	3.20	4.25	5.62	7.42	2.56	3.40	4.50	5.94
2043	2.02	2.71	3.62	4.82	2.48	3.33	4.45	5.93	2.23	2.99	4.00	5.33	3.32	4.45	5.95	7.93	2.63	3.52	4.70	6.27

Table S-4, continued. Projected fuel price indices with assumed general price inflation rates of 2 %, 3 %, 4 %, and 5 %, by end-use sector and fuel type.

Census Region 4 (Alaska, Arizona, California, Colorado, Hawaii, Idaho, Montana, Nevada, New Mexico, Oregon, Utah, Washington, Wyoming)

Projected April 1 Fuel Price Indices (April 1, 2013 = 1.00)

---INDUSTRIAL---

Year	Electricity				Distillate Oil				Residual Oil				Natural Gas				Coal			
	Inflation Rate				Inflation Rate				Inflation Rate				Inflation Rate				Inflation Rate			
	2 %	3 %	4 %	5 %	2 %	3 %	4 %	5 %	2 %	3 %	4 %	5 %	2 %	3 %	4 %	5 %	2 %	3 %	4 %	5 %
2014	1.01	1.02	1.03	1.04	0.92	0.93	0.93	0.94	0.79	0.80	0.81	0.81	1.06	1.08	1.09	1.10	1.02	1.03	1.04	1.05
2015	1.02	1.04	1.06	1.08	0.91	0.93	0.95	0.97	0.74	0.75	0.77	0.78	1.10	1.12	1.14	1.16	1.05	1.07	1.10	1.12
2016	1.06	1.09	1.12	1.15	0.95	0.97	1.00	1.03	0.76	0.78	0.80	0.83	1.20	1.23	1.27	1.31	1.08	1.11	1.15	1.18
2017	1.09	1.13	1.18	1.22	0.99	1.03	1.07	1.11	0.79	0.82	0.86	0.89	1.30	1.35	1.40	1.46	1.10	1.14	1.19	1.23
2018	1.12	1.18	1.23	1.29	1.03	1.08	1.13	1.19	0.83	0.87	0.91	0.96	1.40	1.47	1.55	1.62	1.11	1.17	1.22	1.28
2019	1.13	1.20	1.28	1.35	1.07	1.14	1.20	1.27	0.86	0.92	0.97	1.03	1.49	1.58	1.68	1.78	1.13	1.20	1.27	1.35
2020	1.16	1.24	1.33	1.42	1.11	1.19	1.27	1.36	0.90	0.96	1.03	1.10	1.55	1.66	1.78	1.90	1.16	1.24	1.33	1.42
2021	1.19	1.28	1.39	1.50	1.15	1.25	1.35	1.45	0.94	1.02	1.10	1.18	1.63	1.76	1.90	2.05	1.19	1.29	1.39	1.50
2022	1.21	1.32	1.44	1.57	1.20	1.31	1.43	1.56	0.98	1.07	1.17	1.27	1.71	1.86	2.03	2.22	1.23	1.34	1.46	1.59
2023	1.23	1.36	1.50	1.65	1.25	1.37	1.51	1.67	1.03	1.14	1.25	1.38	1.79	1.97	2.17	2.39	1.26	1.39	1.53	1.69
2024	1.26	1.40	1.56	1.73	1.29	1.44	1.60	1.77	1.08	1.21	1.35	1.50	1.84	2.05	2.28	2.53	1.30	1.45	1.61	1.79
2025	1.29	1.45	1.63	1.83	1.34	1.50	1.69	1.89	1.14	1.28	1.44	1.61	1.87	2.10	2.36	2.65	1.34	1.51	1.69	1.90
2026	1.33	1.51	1.71	1.94	1.38	1.57	1.78	2.02	1.19	1.35	1.53	1.73	1.90	2.15	2.44	2.77	1.39	1.57	1.78	2.02
2027	1.36	1.56	1.78	2.04	1.43	1.64	1.88	2.15	1.24	1.42	1.62	1.86	1.93	2.21	2.53	2.89	1.43	1.64	1.88	2.15
2028	1.38	1.60	1.85	2.14	1.48	1.71	1.98	2.29	1.29	1.49	1.72	1.99	1.98	2.29	2.65	3.06	1.48	1.71	1.98	2.28
2029	1.42	1.66	1.93	2.25	1.53	1.79	2.09	2.43	1.34	1.57	1.83	2.14	2.05	2.39	2.79	3.25	1.52	1.78	2.07	2.42
2030	1.46	1.72	2.03	2.39	1.58	1.87	2.20	2.59	1.40	1.66	1.95	2.30	2.11	2.49	2.93	3.45	1.57	1.86	2.19	2.57
2031	1.50	1.79	2.13	2.53	1.63	1.95	2.32	2.75	1.47	1.75	2.09	2.48	2.18	2.59	3.09	3.67	1.63	1.94	2.31	2.74
2032	1.55	1.86	2.24	2.69	1.69	2.03	2.44	2.93	1.53	1.85	2.22	2.66	2.26	2.72	3.27	3.92	1.68	2.03	2.44	2.92
2033	1.59	1.93	2.35	2.84	1.75	2.12	2.58	3.12	1.59	1.93	2.34	2.84	2.36	2.87	3.48	4.21	1.75	2.12	2.58	3.12
2034	1.64	2.02	2.47	3.02	1.81	2.23	2.73	3.33	1.66	2.04	2.50	3.05	2.47	3.03	3.71	4.54	1.81	2.22	2.72	3.32
2035	1.70	2.11	2.61	3.23	1.89	2.34	2.89	3.57	1.74	2.16	2.67	3.30	2.60	3.23	3.99	4.93	1.87	2.31	2.86	3.53
2036	1.77	2.22	2.77	3.45	1.96	2.46	3.07	3.82	1.80	2.26	2.82	3.51	2.76	3.45	4.31	5.38	1.93	2.42	3.02	3.76
2037	1.84	2.33	2.93	3.69	2.04	2.58	3.25	4.09	1.90	2.40	3.02	3.80	2.93	3.70	4.67	5.88	1.98	2.51	3.16	3.98
2038	1.92	2.45	3.11	3.96	2.11	2.69	3.43	4.35	1.99	2.54	3.23	4.10	3.11	3.97	5.05	6.42	2.04	2.60	3.31	4.21
2039	1.99	2.57	3.30	4.23	2.19	2.82	3.62	4.65	2.08	2.68	3.44	4.41	3.25	4.19	5.39	6.91	2.11	2.72	3.49	4.48
2040	2.07	2.70	3.50	4.54	2.27	2.95	3.83	4.96	2.16	2.81	3.65	4.72	3.40	4.42	5.74	7.44	2.17	2.82	3.66	4.74
2041	2.14	2.81	3.69	4.82	2.35	3.09	4.06	5.30	2.26	2.96	3.88	5.08	3.55	4.66	6.11	7.99	2.23	2.93	3.85	5.03
2042	2.20	2.92	3.86	5.10	2.44	3.24	4.29	5.67	2.36	3.13	4.15	5.47	3.71	4.92	6.51	8.59	2.29	3.05	4.03	5.32
2043	2.26	3.03	4.05	5.40	2.54	3.40	4.55	6.06	2.47	3.31	4.43	5.90	3.87	5.19	6.93	9.24	2.36	3.16	4.22	5.63

Table S-5. Projected fuel price indices with assumed general price inflation rates of 2 %, 3 %, 4 %, and 5 %, by end-use sector and fuel type.

United States Average

Projected April 1 Fuel Price Indices (April 1, 2013 = 1.00)

---------RESIDENTIAL---------

Year	Electricity				Distillate Oil				LPG				Natural Gas			
	Inflation Rate				Inflation Rate				Inflation Rate				Inflation Rate			
	2 %	3 %	4 %	5 %	2 %	3 %	4 %	5 %	2 %	3 %	4 %	5 %	2 %	3 %	4 %	5 %
2014	1.02	1.04	1.05	1.06	0.99	1.00	1.01	1.02	0.99	1.00	1.01	1.02	1.00	1.01	1.02	1.03
2015	1.05	1.07	1.09	1.11	1.01	1.03	1.05	1.07	0.98	1.00	1.02	1.04	1.01	1.03	1.05	1.07
2016	1.08	1.12	1.15	1.18	1.04	1.07	1.10	1.13	1.00	1.03	1.06	1.09	1.07	1.10	1.13	1.16
2017	1.12	1.16	1.21	1.25	1.08	1.12	1.17	1.21	1.04	1.09	1.13	1.17	1.13	1.17	1.22	1.27
2018	1.14	1.20	1.26	1.32	1.12	1.17	1.23	1.29	1.09	1.14	1.20	1.26	1.19	1.25	1.31	1.38
2019	1.16	1.23	1.30	1.38	1.16	1.23	1.30	1.37	1.13	1.20	1.27	1.35	1.24	1.32	1.40	1.48
2020	1.18	1.26	1.35	1.45	1.20	1.28	1.37	1.46	1.18	1.26	1.35	1.44	1.29	1.38	1.48	1.58
2021	1.20	1.30	1.41	1.52	1.24	1.34	1.45	1.56	1.22	1.31	1.42	1.53	1.33	1.44	1.56	1.68
2022	1.23	1.34	1.46	1.60	1.28	1.40	1.53	1.66	1.26	1.37	1.50	1.63	1.39	1.51	1.65	1.80
2023	1.26	1.38	1.52	1.68	1.33	1.46	1.61	1.77	1.30	1.43	1.57	1.73	1.44	1.59	1.75	1.92
2024	1.28	1.43	1.59	1.77	1.38	1.53	1.70	1.89	1.34	1.49	1.66	1.84	1.49	1.66	1.84	2.05
2025	1.31	1.48	1.66	1.86	1.43	1.60	1.80	2.02	1.38	1.55	1.74	1.95	1.53	1.72	1.93	2.17
2026	1.35	1.53	1.73	1.96	1.48	1.68	1.90	2.15	1.42	1.61	1.82	2.07	1.58	1.80	2.04	2.31
2027	1.38	1.58	1.81	2.07	1.53	1.75	2.01	2.29	1.46	1.67	1.91	2.19	1.63	1.87	2.14	2.44
2028	1.41	1.64	1.89	2.18	1.58	1.83	2.12	2.44	1.50	1.73	2.00	2.31	1.68	1.94	2.24	2.59
2029	1.45	1.69	1.97	2.30	1.64	1.91	2.24	2.60	1.54	1.80	2.10	2.45	1.73	2.02	2.36	2.75
2030	1.48	1.74	2.06	2.42	1.69	2.00	2.36	2.77	1.58	1.86	2.20	2.59	1.78	2.10	2.48	2.92
2031	1.51	1.80	2.15	2.55	1.75	2.09	2.48	2.95	1.62	1.93	2.30	2.73	1.84	2.20	2.61	3.10
2032	1.55	1.86	2.24	2.69	1.81	2.18	2.62	3.14	1.66	2.00	2.41	2.89	1.90	2.29	2.75	3.30
2033	1.58	1.93	2.34	2.83	1.87	2.28	2.76	3.35	1.71	2.08	2.52	3.05	1.97	2.39	2.90	3.52
2034	1.62	1.99	2.44	2.98	1.94	2.38	2.92	3.57	1.76	2.16	2.64	3.23	2.05	2.52	3.08	3.77
2035	1.67	2.07	2.56	3.16	2.01	2.50	3.09	3.81	1.81	2.24	2.77	3.42	2.14	2.66	3.28	4.05
2036	1.72	2.15	2.69	3.35	2.09	2.62	3.27	4.07	1.86	2.33	2.91	3.63	2.25	2.81	3.51	4.37
2037	1.77	2.24	2.83	3.56	2.17	2.74	3.46	4.35	1.92	2.43	3.06	3.85	2.35	2.97	3.75	4.72
2038	1.83	2.33	2.97	3.77	2.24	2.86	3.65	4.63	1.98	2.52	3.21	4.08	2.46	3.15	4.01	5.09
2039	1.88	2.43	3.12	4.00	2.33	3.00	3.86	4.94	2.04	2.62	3.37	4.33	2.56	3.30	4.24	5.43
2040	1.93	2.51	3.26	4.23	2.41	3.14	4.07	5.28	2.09	2.72	3.54	4.58	2.65	3.45	4.48	5.81
2041	1.98	2.60	3.41	4.46	2.50	3.29	4.31	5.63	2.16	2.83	3.72	4.86	2.75	3.61	4.74	6.19
2042	2.03	2.69	3.56	4.70	2.59	3.44	4.56	6.01	2.22	2.95	3.90	5.15	2.85	3.78	5.00	6.60
2043	2.08	2.79	3.72	4.96	2.69	3.61	4.82	6.42	2.29	3.07	4.10	5.46	2.95	3.95	5.28	7.03

Table S-5, continued. Projected fuel price indices with assumed general price inflation rates of 2 %, 3 %, 4 %, and 5 %, by end-use sector and fuel type.

United States Average

Projected April 1 Fuel Price Indices (April 1, 2013 = 1.00)

----COMMERCIAL----

Year	Electricity				Distillate Oil				Residual Oil				Natural Gas				Coal			
	Inflation Rate				Inflation Rate				Inflation Rate				Inflation Rate				Inflation Rate			
	2 %	3 %	4 %	5 %	2 %	3 %	4 %	5 %	2 %	3 %	4 %	5 %	2 %	3 %	4 %	5 %	2 %	3 %	4 %	5 %
2014	1.02	1.03	1.04	1.05	0.92	0.93	0.94	0.95	0.77	0.78	0.79	0.80	1.01	1.02	1.03	1.04	1.01	1.02	1.03	1.04
2015	1.03	1.05	1.07	1.10	0.91	0.93	0.95	0.97	0.72	0.74	0.75	0.76	1.01	1.03	1.05	1.07	1.05	1.07	1.09	1.11
2016	1.07	1.10	1.13	1.16	0.94	0.97	1.00	1.03	0.75	0.77	0.79	0.82	1.07	1.10	1.14	1.17	1.08	1.11	1.15	1.18
2017	1.10	1.15	1.19	1.24	0.98	1.02	1.06	1.10	0.77	0.81	0.84	0.87	1.13	1.18	1.22	1.27	1.11	1.15	1.20	1.25
2018	1.13	1.19	1.25	1.31	1.02	1.07	1.13	1.18	0.81	0.85	0.90	0.94	1.20	1.26	1.32	1.39	1.14	1.20	1.26	1.32
2019	1.15	1.22	1.29	1.37	1.06	1.13	1.19	1.26	0.85	0.90	0.95	1.01	1.25	1.33	1.41	1.49	1.18	1.25	1.32	1.40
2020	1.17	1.26	1.35	1.44	1.10	1.18	1.26	1.35	0.88	0.95	1.01	1.08	1.30	1.39	1.49	1.59	1.21	1.29	1.38	1.48
2021	1.20	1.30	1.40	1.51	1.14	1.23	1.33	1.44	0.92	1.00	1.08	1.16	1.34	1.45	1.57	1.69	1.24	1.34	1.45	1.56
2022	1.23	1.34	1.46	1.59	1.19	1.30	1.41	1.54	0.96	1.05	1.15	1.25	1.40	1.52	1.66	1.81	1.27	1.39	1.51	1.65
2023	1.25	1.38	1.52	1.67	1.23	1.36	1.50	1.65	1.01	1.11	1.22	1.34	1.45	1.60	1.76	1.94	1.31	1.44	1.58	1.74
2024	1.27	1.42	1.58	1.75	1.28	1.42	1.58	1.76	1.05	1.17	1.30	1.45	1.50	1.67	1.86	2.07	1.34	1.49	1.66	1.85
2025	1.29	1.46	1.63	1.83	1.33	1.49	1.67	1.88	1.09	1.23	1.38	1.55	1.54	1.73	1.95	2.18	1.38	1.55	1.74	1.95
2026	1.32	1.49	1.69	1.92	1.37	1.56	1.77	2.00	1.14	1.30	1.47	1.67	1.59	1.81	2.05	2.32	1.42	1.61	1.82	2.06
2027	1.34	1.54	1.76	2.01	1.42	1.63	1.87	2.13	1.20	1.37	1.57	1.79	1.64	1.88	2.15	2.46	1.46	1.67	1.91	2.18
2028	1.37	1.59	1.83	2.12	1.47	1.70	1.97	2.27	1.25	1.45	1.67	1.93	1.69	1.95	2.26	2.60	1.50	1.73	2.00	2.31
2029	1.40	1.64	1.91	2.23	1.52	1.78	2.08	2.42	1.30	1.52	1.78	2.07	1.74	2.03	2.37	2.76	1.54	1.80	2.10	2.45
2030	1.43	1.69	2.00	2.35	1.57	1.86	2.19	2.58	1.36	1.61	1.90	2.23	1.79	2.11	2.49	2.93	1.59	1.88	2.21	2.60
2031	1.47	1.75	2.08	2.48	1.63	1.94	2.31	2.74	1.40	1.67	1.99	2.36	1.85	2.20	2.62	3.11	1.63	1.95	2.32	2.75
2032	1.51	1.81	2.18	2.61	1.68	2.03	2.44	2.92	1.47	1.77	2.12	2.55	1.90	2.29	2.75	3.30	1.68	2.02	2.43	2.91
2033	1.54	1.88	2.28	2.76	1.75	2.12	2.57	3.12	1.53	1.86	2.26	2.74	1.97	2.39	2.90	3.52	1.73	2.10	2.55	3.09
2034	1.59	1.95	2.38	2.91	1.81	2.22	2.72	3.33	1.59	1.95	2.39	2.92	2.05	2.52	3.08	3.77	1.78	2.18	2.68	3.27
2035	1.64	2.03	2.51	3.09	1.88	2.33	2.89	3.56	1.67	2.07	2.56	3.16	2.15	2.66	3.29	4.06	1.84	2.28	2.81	3.47
2036	1.69	2.12	2.65	3.30	1.96	2.45	3.06	3.81	1.75	2.19	2.73	3.41	2.26	2.82	3.53	4.39	1.89	2.37	2.96	3.69
2037	1.75	2.21	2.79	3.51	2.04	2.57	3.25	4.08	1.83	2.31	2.91	3.67	2.37	2.99	3.77	4.75	1.95	2.46	3.11	3.91
2038	1.81	2.31	2.94	3.74	2.11	2.69	3.42	4.35	1.90	2.42	3.08	3.91	2.49	3.18	4.04	5.14	2.01	2.56	3.26	4.14
2039	1.88	2.42	3.11	3.99	2.19	2.82	3.62	4.65	1.99	2.56	3.29	4.22	2.58	3.33	4.28	5.49	2.07	2.67	3.43	4.40
2040	1.94	2.52	3.27	4.23	2.27	2.95	3.84	4.97	2.08	2.70	3.51	4.54	2.68	3.49	4.53	5.87	2.13	2.77	3.60	4.66
2041	1.99	2.61	3.42	4.47	2.36	3.10	4.06	5.31	2.17	2.85	3.74	4.89	2.79	3.66	4.80	6.28	2.19	2.88	3.77	4.93
2042	2.04	2.70	3.57	4.72	2.45	3.25	4.30	5.68	2.27	3.02	3.99	5.27	2.90	3.84	5.09	6.72	2.25	2.98	3.95	5.21
2043	2.09	2.80	3.74	4.98	2.54	3.41	4.56	6.07	2.38	3.19	4.26	5.68	3.01	4.03	5.39	7.18	2.31	3.09	4.13	5.51

Table S-5, continued. Projected fuel price indices with assumed general price inflation rates of 2 %, 3 %, 4 %, and 5 %, by end-use sector and fuel type.

United States Average

Projected April 1 Fuel Price Indices (April 1, 2013 = 1.00)

-----INDUSTRIAL-----

Year	Electricity 2 %	3 %	4 %	5 %	Distillate Oil 2 %	3 %	4 %	5 %	Residual Oil 2 %	3 %	4 %	5 %	Natural Gas 2 %	3 %	4 %	5 %	Coal 2 %	3 %	4 %	5 %
2014	1.01	1.02	1.03	1.04	0.91	0.92	0.93	0.94	0.89	0.89	0.90	0.91	1.07	1.08	1.09	1.10	1.02	1.03	1.04	1.05
2015	1.02	1.04	1.06	1.08	0.90	0.92	0.94	0.96	0.86	0.88	0.90	0.91	1.10	1.12	1.14	1.16	1.05	1.07	1.09	1.12
2016	1.06	1.10	1.13	1.16	0.93	0.96	0.99	1.02	0.89	0.91	0.94	0.97	1.21	1.24	1.28	1.32	1.09	1.12	1.15	1.19
2017	1.10	1.15	1.19	1.24	0.97	1.01	1.05	1.09	0.92	0.96	0.99	1.03	1.30	1.35	1.40	1.46	1.11	1.16	1.21	1.25
2018	1.14	1.19	1.25	1.31	1.01	1.06	1.12	1.17	0.96	1.01	1.06	1.11	1.39	1.46	1.53	1.61	1.14	1.20	1.26	1.32
2019	1.16	1.23	1.30	1.38	1.05	1.12	1.18	1.25	1.00	1.06	1.12	1.19	1.46	1.55	1.64	1.74	1.18	1.25	1.32	1.40
2020	1.18	1.26	1.35	1.45	1.09	1.17	1.25	1.34	1.04	1.11	1.19	1.27	1.51	1.62	1.73	1.85	1.20	1.29	1.38	1.48
2021	1.21	1.31	1.41	1.53	1.13	1.23	1.33	1.43	1.08	1.17	1.26	1.36	1.57	1.70	1.84	1.98	1.24	1.34	1.44	1.56
2022	1.24	1.36	1.48	1.61	1.18	1.29	1.41	1.54	1.12	1.23	1.34	1.46	1.65	1.80	1.97	2.15	1.27	1.38	1.51	1.65
2023	1.28	1.41	1.55	1.70	1.23	1.35	1.49	1.64	1.17	1.29	1.42	1.56	1.74	1.92	2.11	2.32	1.30	1.44	1.58	1.74
2024	1.31	1.45	1.62	1.80	1.27	1.42	1.57	1.75	1.22	1.36	1.51	1.68	1.81	2.01	2.24	2.49	1.34	1.49	1.66	1.85
2025	1.34	1.50	1.69	1.89	1.32	1.48	1.66	1.87	1.27	1.42	1.60	1.79	1.86	2.09	2.35	2.64	1.38	1.55	1.74	1.95
2026	1.37	1.55	1.76	1.99	1.36	1.55	1.76	1.99	1.32	1.50	1.70	1.92	1.93	2.19	2.48	2.81	1.42	1.61	1.82	2.06
2027	1.40	1.61	1.84	2.10	1.41	1.62	1.85	2.12	1.37	1.57	1.80	2.06	1.98	2.27	2.60	2.97	1.46	1.67	1.91	2.19
2028	1.44	1.66	1.92	2.22	1.46	1.69	1.95	2.25	1.43	1.66	1.92	2.21	2.05	2.37	2.74	3.17	1.49	1.73	2.00	2.31
2029	1.48	1.72	2.01	2.35	1.51	1.76	2.06	2.40	1.49	1.74	2.03	2.37	2.12	2.48	2.90	3.38	1.54	1.80	2.10	2.44
2030	1.51	1.79	2.11	2.48	1.56	1.84	2.17	2.55	1.55	1.83	2.16	2.54	2.19	2.59	3.05	3.59	1.58	1.87	2.20	2.59
2031	1.56	1.85	2.21	2.62	1.61	1.92	2.28	2.71	1.60	1.91	2.27	2.70	2.27	2.71	3.22	3.83	1.63	1.94	2.31	2.75
2032	1.60	1.92	2.31	2.77	1.66	2.00	2.41	2.89	1.67	2.01	2.41	2.89	2.35	2.83	3.40	4.08	1.68	2.02	2.42	2.91
2033	1.65	2.00	2.43	2.94	1.72	2.10	2.54	3.08	1.74	2.11	2.56	3.10	2.45	2.97	3.61	4.37	1.72	2.09	2.54	3.08
2034	1.70	2.08	2.55	3.12	1.79	2.20	2.69	3.29	1.80	2.21	2.71	3.31	2.58	3.16	3.88	4.74	1.77	2.17	2.66	3.25
2035	1.76	2.18	2.70	3.33	1.86	2.31	2.85	3.52	1.89	2.34	2.89	3.57	2.73	3.38	4.18	5.16	1.83	2.27	2.80	3.46
2036	1.83	2.29	2.86	3.57	1.94	2.42	3.03	3.77	1.97	2.46	3.08	3.83	2.91	3.64	4.55	5.67	1.89	2.36	2.95	3.67
2037	1.91	2.41	3.04	3.82	2.01	2.54	3.21	4.04	2.05	2.59	3.27	4.12	3.10	3.92	4.94	6.22	1.94	2.45	3.09	3.89
2038	1.98	2.53	3.22	4.09	2.08	2.66	3.39	4.30	2.13	2.72	3.46	4.39	3.30	4.21	5.36	6.81	2.00	2.55	3.24	4.12
2039	2.06	2.65	3.41	4.37	2.16	2.79	3.58	4.59	2.22	2.86	3.68	4.72	3.44	4.44	5.70	7.32	2.05	2.64	3.40	4.36
2040	2.13	2.77	3.59	4.65	2.24	2.92	3.79	4.91	2.32	3.01	3.91	5.06	3.60	4.69	6.08	7.88	2.11	2.75	3.57	4.62
2041	2.19	2.88	3.77	4.93	2.33	3.06	4.01	5.25	2.42	3.17	4.16	5.44	3.77	4.96	6.50	8.50	2.17	2.85	3.74	4.89
2042	2.25	2.99	3.96	5.22	2.42	3.21	4.25	5.61	2.52	3.35	4.43	5.85	3.96	5.25	6.95	9.17	2.23	2.96	3.92	5.17
2043	2.32	3.11	4.15	5.53	2.51	3.37	4.50	6.00	2.64	3.53	4.72	6.29	4.14	5.55	7.42	9.88	2.30	3.08	4.12	5.49